TERRORISM
AND THE PRESS

David Copeland
General Editor

Vol. 3

PETER LANG
New York • Washington, D.C./Baltimore • Bern
Frankfurt am Main • Berlin • Brussels • Vienna • Oxford

Brooke Barnett & Amy Reynolds

TERRORISM
AND THE PRESS

AN UNEASY RELATIONSHIP

PETER LANG
New York • Washington, D.C./Baltimore • Bern
Frankfurt am Main • Berlin • Brussels • Vienna • Oxford

Library of Congress Cataloging-in-Publication Data

Barnett, Brooke.
Terrorism and the press: an uneasy relationship / Brooke Barnett, Amy Reynolds.
p. cm. — (Mediating American history; vol. 3)
Includes bibliographical references and index.
1. Terrorism—Press coverage. I. Reynolds, Amy. II. Title.
PN4784.T45B37   172'.1—dc22   2008033475
ISBN 978-0-8204-9517-0 (hardcover)
ISBN 978-0-8204-9516-3 (paperback)
ISSN 0085-2473

Bibliographic information published by **Die Deutsche Nationalbibliothek**.
Die Deutsche Nationalbibliothek lists this publication in the "Deutsche
Nationalbibliografie"; detailed bibliographic data are available
on the Internet at http://dnb.d-nb.de/.

Cover design by Ken Calhoun

The paper in this book meets the guidelines for permanence and durability
of the Committee on Production Guidelines for Book Longevity
of the Council of Library Resources.

Printed in the United States of America

For Tom, Lily, and Jack
—BROOKE BARNETT

For Tad, Chase, Mom, and Dad
—AMY REYNOLDS

CONTENTS

ACKNOWLEDGMENTS

Since 2001, we have collaborated on many projects and have had the great benefit of each other's creativity, expertise, and, most importantly, friendship. This book is no exception. Many others have helped us bring this project to life. We would like to thank four of our students and graduates from Elon University and Indiana University who provided research assistance—Allison Deiboldt, Tayler Kent, Cristina Mayer, and John Seasly. Our thanks go to Laura Roselle and Sarah Oates for their contributions to a study that forms the original research in Chapter 6. We are fortunate to have as our series editor David Copeland, a kind, generous, and gifted colleague. Mary Savigar, Acquisitions Editor: Media and Communication Studies at Peter Lang Publishing, has been an efficient and professional editor and a joy to work with from start to finish. We are also grateful to Sophie Appel, production supervisor at Peter Lang, for the terrific work she does and for her help in navigating the production process. We would also like to thank Fred H. Cate, Brad Hamm, Lesa Major, Connie Book, Barbara Miller and Tom Mould for their friendship and collegiality. We not only benefit from the support of our colleagues, but also from our respective universities, Elon University and Indiana University, that provide us a terrific environment in which to pursue our research endeavors. Finally, we are indebted to our families, who have waited patiently for us "to finish this one last thing" before we spend time with them.

INTRODUCTION

Federal Aviation Administration: Hi. Boston Center TMU [Traffic Management Unit], we have a problem here. We have a hijacked aircraft headed towards New York, and we need you guys to, we need someone to scramble some F-16s or something up there, help us out.

North East Air Defense Center: Is this real-world or exercise?

Federal Aviation Administration: No, this is not an exercise, not a test.
—The 9/11 Commission Report, detailing initial military notification
and response on September 11, 2001[1]

On a clear day in September 2001, the U.S. networks were wrapping up their morning news shows. Matt Lauer of NBC's *Today* show interrupted his interview with an author to report an incident at the World Trade Center. This began the media coverage of the deadliest terrorist attack in U.S. history. At the end of the day, four airplanes had been hijacked and almost three thousand people were dead in New York at the World Trade Center Towers; in Washington, DC, at the Pentagon; and a remote site in rural Pennsylvania where the fourth plane crashed.

Although many terrorists have gained international media attention for their attacks during the past fifty years, the attacks of September 11 received unprecedented international news media coverage. Terrorism has often been likened to theater, but now it seems that connecting terrorism to the mass media, whose attention it still holds as a result of September 11, is more appropriate. Breaking news is the new vehicle to transport the terrorist message. The September 11 attacks spawned five days of commercial-free live breaking news in the United States alone, screen time valued in the millions.[2] Even sports and entertainment channels covered the crisis or altered their programming. The FOX sports channel simply aired an image of the U.S. flag. Both *Newsweek* and *Time* magazine featured September 11–related images on the cover of their next eight issues. Droves of famous people came out for a televised charity event to raise money for victims. Fiction films that included terrorist themes or even backdrops of the Twin Towers were delayed or never released. The story dominated every aspect of American media.

Polls showed that the American public was paying attention. Essentially, all Americans followed the news of the attacks on either the radio or television, creating an unprecedented audience, ready to receive information from the official government sources whose presence dominated the early coverage. Terrorism remained on the public and media agenda, with the media focused on terrorism and more than 90 percent of Americans following it closely for the next six weeks.[3]

Even when regularly scheduled programming resumed, it was not the same. Blurring the line between fact and fiction, the NBC drama about the American presidency, The West Wing, scrapped its planned show and aired an episode dealing with a terrorist attack. Late Night talk show host David Letterman went back on air with no attempts at humor. His guest, CBS news anchor Dan Rather, who in the past cracked jokes and sang songs on the show, was moved to tears twice. Even NBC's improvisational comedy show Saturday Night Live and Comedy Central's The Daily Show with Jon Stewart seemed to cut government officials some slack; their political humor was less pointed. The one comedian who deviated from this unspoken code received a swift punishment. When Bill Maher, host of the ABC late night show Politically Incorrect said that the terrorists were not cowards because they sacrificed their lives for their cause, but that perhaps the U.S. military practice of lobbing cruise missiles from two thousand miles away was cowardly, public outcry ensued, and his show was eventually cancelled.

The terrorists responsible for the September 11 attacks had commanded the attention of a nation and its leaders through use of the target nation's own media, news, comedy shows, and dramas. The coverage of the September 11 attacks reaffirms what some political scientists have said for years—that the media are a central player in terrorist activity.[4] Former British Prime Minister Margaret Thatcher, when dealing with threats from the Irish Republican Army, said that publicity is the oxygen of terrorism. According to political scientist Brigette Nacos, "if anything has changed in the last ten or fifteen years, it is the increased availability of the sort of oxygen Mrs. Thatcher warned of and upon which mass-mediated terrorism thrives."[5]

Clearly, the news media are an integral piece of the contemporary terrorists' calculations. As the U.S. State Department definition states, terrorism is "premeditated, politically motivated violence perpetrated against noncombatant targets by subnational groups or clandestine state agents, usually intended to influence an audience."[6] Today's media can deliver a larger global audience than ever before, usually in nearly real time. The impact of this can't be understated—the attacks of September 11 led to the U.S. wars in Afghanistan and Iraq, to the largest restructuring of the U.S. federal government since the creation of the Department of Defense, and to the passage of legislation that significantly impacts the civil liberties of all Americans. Terrorism experts concur that the terrorist act is not usually about the victims but rather about terrorists winning the attention of the news media, the public, and the government. As coverage of September 11 showed, media are delivering the terrorist's message in nearly every conceivable way.

Many books have been written about the press and terrorism, particularly since September 11. This book differs in that it is a press-focused exploration of the relationship between the media and terrorism. This book explores terrorism through the lens of media theories and with a strong consideration of the realities and pressures of daily journalism. People from many different disciplines study terrorism. Some scholars bring the press into the study of terrorism as a variable in a political science context. Other scholars examine terrorism from a sociological perspective, often connecting to media without a clear understanding of the important role the press plays during times of crisis. Much research outside of the field of mass communication dismisses or ignores the constraints placed on journalists who make up the press, and in their analyses these scholars often suggest ways for government to circumvent the press as a solution to the potential power the media can bring to a terrorist. While arguments in favor of press censorship have existed for years, most of the time these kinds of recommendations are not legally or socially reconcilable in a democracy with First Amendment protections for press.

When a scholar embarks on any kind of research, he or she must consider the goal of the research. In most of the research on terrorism that comes from other disciplines (criminal justice, military science, and political science, for example), the goal is often to make governmental policy recommendations or to figure out targeted responses that will help a government fight terrorism. Some of these recommendations include condemnation of the news media that often come without a clear understanding of how the media work.

One value of studying terrorism through a news media lens is not to develop policy, but rather to promote better interdisciplinary understanding. Previous research tells us that the media are the primary vehicles through which a significant number of people learn about and come to understand terrorism. This should lead mass communication scholars to consider how the press reports on terrorism, how that reporting varies depending on the medium, and globally and within different media and government systems. How important is the press to spreading fear? Does the press help terrorists recruit? Is the press wrongly giving voice to the political grievances of murderers by communicating their actions to a wider audience? These are some of the issues this book addresses, through the application of a variety of mass communication theories and methods.

Within the field of journalism and mass communication, no other scholars have explored the relationship between the press and terrorism in such a focused, yet broad and wide-reaching way. Typically, as one will see cited throughout this book, studies about media coverage of terrorism treat terrorism as an isolated issue or event, as if it is similar to reporting on a political campaign or on an earthquake. Our colleagues in other disciplines who have studied the complexities of terrorism show us that this kind of approach is short-sighted. Other accounts that explore the media and terrorism, particularly those connected to September 11, are firsthand anecdotal stories written by practicing journalists. All of these are useful and interesting, yet none is complete. These accounts also are absent theory to help put them into a broader, more useful context.

In addition to using a press-centered approach to explore the news media and its relationship to terrorism, this book also seeks to highlight terrorism-related research within the field of journalism and mass communication; expand on the terrorism-related research from other fields to incorporate the press perspective; and to provide original research that better explains how the news media view the so-called symbiotic relationship between the media and terrorism.[7]

One idea that generates agreement among all who study the press and terrorism is that the way that terrorism is framed dictates the ways that the public may perceive it. The relationship between the media and the government also dictates the type of coverage. Some experts argue that media coverage lends credibility to the terrorist and encourages further attacks. Other experts say that the media rely too heavily on governmental interpretation of events and elite sources and simply allow the news to become the broadcast or reprint of a government press release. A recent study by economists showed that coverage of terrorism caused more attacks, and more terrorism attacks caused more coverage—a mutually beneficial spiral of death that the authors say has increased because of a heightened interest in terrorism since September 11, 2001.[8]

Even the untutored can state the adage "One person's terrorist is another's freedom fighter." But this adage gets us no further in parsing out what is meant by the word "terrorist" when used by the press, the government, and an individual in regular conversation. In fact, many accuse the United States of only using the term "terrorist" for anti-Western activity. Use of the word terrorism in media coverage often conforms to what the government calls terrorism. Even definitions of terrorism among U.S. federal government agencies vary. The Federal Bureau of Investigation, the Department of State, and federal and criminal procedures provide different definitions of the term "terrorism." The United Nations has been unable to agree on a definition of terrorism. Chapter 1 explores these definitional issues, as well as the history of terrorism.

The press has an equally hard time deciding when to call something terrorism. The British Broadcasting Corporation (BBC) tries not to use the word terrorism in its reporting at all. The network instead uses words like "attack" and "bombing." According to BBC guidelines, the United States was attacked on September 11, and the United States attacked Afghanistan in retaliation. This refusal to use the word terrorist came under fire by U.S. media watchdog groups. Chapter 2 explores how the press uses the term, and it lays the theoretical foundation for examining terrorism from a press perspective.

Agenda setting is a mass communication theory that holds that the media do not tell you what to think, but they do tell you what to think about. The gatekeepers in news, those who make story assignments and decide what stories are covered and how, are helping to decide if the public is going to be talking about terrorism. Sometimes the way gatekeepers talk about an issue determines how a story is framed. Framing is defined as "a central organizing idea for news content that supplies a context and suggests what the issue is through the use of selection, emphasis, exclusion and elaboration."[9] Terrorism is framed differently depending on who is covering the story, where the story takes place, who the terrorist is, and who the victims are. The demands and nationality of the terrorist, for example, can impact how terrorist activity is framed. Chapter 2 will outline how these theoretical perspectives help us to make sense of the uneasy relationship between the press and the terrorist.

Chapter 3 explores the relationship between the media and the government. Just as terrorists use the media as a platform, the government also uses the press to reach a large audience. After the September 11 attacks, the U.S.

media followed every move of President George W. Bush. Every time the president spoke at "ground zero" or the Pentagon, he received prominent media attention. Sometimes the government will also unintentionally or deliberately steer the press in the wrong direction. If the media are not diligent or informed enough about the issues, they will not prevent this from happening.

An example of the unintentional misinforming of the press came from the 2001 anthrax scares in the United States. The American news media erroneously referred to anthrax as a virus, and often this was because they were taking their cues from ill-informed government officials who do political, not medical, work.[10] The verbatim reporting of information from press conferences left the impression that anthrax was a mutating germ, and made it seem like it was fairly easy to catch it from someone. Although lots of good, in-depth reporting followed these early reports on the anthrax scare, researchers who studied the media coverage argued that once scientific misinformation is out there it is not replaced by corrective information bur rather coexists with it. Often, the headlines are read by the audience (in this case, the headlines are where the misinformation was found), and the in-depth stories that follow are not (the correct information was reported in the stories). Also, reporters draw on past reports for background information and often end up reporting the errors again. The last problem with the reporting on anthrax was the emphasis on the possible use of it as a bioterror. Although this is theoretically possible, it is not probable. The media turned this into an unnecessary scare because of its heavy reliance on uninformed government sources.

The media can also misquote the government and add legitimacy to their own theories about the identity of terrorists. An example comes from CNN's initial coverage of the Okalahoma City bombing in 1995:

> Natalie Allen (CNN anchor): (There's been speculation) about some Islamic groups that are in that city (Oklahoma) that have caused concerns and have been watched. What do you know about that?

> Sen. James Inhofe (R-OK): I think it's probably doing a disservice to be coming to conclusions concerning the Nation of Islam at this point (CNN transcript.)

A few minutes later, in an interview with CNN White House correspondent Jill Dougherty, Allen ignored Senator Inhofe's caution to refrain from assigning blame too soon:

Natalie Allen (CNN anchor): Jill, we were talking with Senator Inhofe of Oklahoma ... He referred to a call that came in from the Nation of Islam and said that he is not contributing any weight to that call as far as any responsibility. *Have you heard anything regarding this,* or is this something the White House is just looking at?

Jill Dougherty (CNN White House correspondent): Natalie, at this point the White House is not going to say much of anything about any type of speculation. They're being extraordinarily careful.

Chapter 3 also examines the Patriot Act and the Office of Homeland Security as well as general access to information issues after September 11. It will also look more generally at access to military operations and war coverage.

The Defense Department policy on combat coverage in the 1980s and 1990s involved several tactics for controlling information. One tactic was to prevent the media from entering the field, which stopped them from getting information. A second tactic was regularly televised White House and Pentagon briefings that took the information directly to the people. The government argued that this would prevent mishandling of the information; the media argued that this was a way to put the official government spin on the information.

Chapter 4 moves the exploration to the images shown in the press coverage of terrorism. Research into the power of images shows that images help people remember an event, likely because they provide an added emotional cue to the information presented. Many current events are now locked in our memory because we have television or still photography memories of those events. The images from September 11 are discussed in detail in chapter 4 to illustrate how the image can suggest meaning beyond what is being said. The visual framing of the September 11 coverage suggests that the press adopted the government's framing of the attacks and that framing became part of the collective memory of the nation on that day. Research has shown that images that are radically different from what we would expect require further information processing, and thus novel images compel viewing and are remembered more readily. An intuitive example of this is the attacks on September 11. The most memorable attacks are those on the World Trade Center towers, likely because we have such graphic images of the two planes hitting the towers, followed by the subsequent tower collapse. This was something that viewers had never seen before, and so it was easily locked into long-term memory. Research has shown that images that are radically different from what we would expect require further information processing and thus novel images compel viewing and are remembered more readily.

Chapter 5 focuses attention on the key role that television plays in terrorism coverage. Coverage of the September 11 attacks, the Okalahoma City bombing, the subway and bus bombings in London, and the train bombings in India will all be discussed. These were all major terrorist events covered extensively by CNN and for which transcripts were available.

Chapter 6 examines the different ways that terrorism is covered at home and abroad with an analysis of British and U.S. coverage of the July 7, 2005, London subway bombings and the September 11, 2001, attacks. This chapter explores how coverage of terrorism varies by event, but also by the country reporting the attack and where the attack takes place. Prior to the terrorism attacks in the 1990s, many Americans thought terrorism was only perpetrated by foreign hands. When the Oklahoma City bombing occurred in the United States in 1995, CNN called it an "event with international implications." Reporters from CNN referred multiple times to two Middle Eastern men in a pickup leaving the scene of the bombing, even after that "rumor" had been proven untrue by the FBI. Also, about three hours after the bombing, one of the Oklahoma City television affiliates received an anonymous phone call from a man claiming to be affiliated with a radical Islamic group that claimed responsibility for the bombing. CNN's affiliate broadcast that information immediately, which fueled speculation that the incident involved Middle Eastern terrorists:

> Local Anchor: (joined in progress) ... Islamic militantism, and we certainly would be doing a disservice to the Muslim Islamic community if we were to assume that at this point, so please take that in spirit, that it *is an unconfirmed phone call, an anonymous phone call* that we received earlier.

Another example comes from CNN anchor Reid Collins:

> Any connection with that, any connection with Sheik Rahman, who's being held in connection with the World Trade Center bombing, or Yousef, who was recently apprehended overseas–these are all potential leads, potential connections that law enforcement officials are looking at very, very vigorously ... There has been some concern that I've heard for some time, about the overall possibility, vulnerability of something like this in an open society, and there's a great deal of work to be done now.

Chapter 7 discusses how the public sends a confusing message when it comes to a patriotic press. Surveys show that a majority of the U.S. public wants a watchdog press and yet desires and expects a pro-American slant to the news. For journalists covering terrorism, this may be an untenable combination. For example, the journalists on September 11 were, like many Americans, shocked at the attacks that day, but they were required to remain on air, mourning, while still continuing to do their jobs. It is perhaps not surprising then that many of them made personal comments that they normally would not feel comfortable making. Meanwhile, studies show that support for a critical press drops after a crisis. So, journalists reporting in the days that followed September 11 faced a public expecting a patriotic press. Journalists who did not deliver faced harsh repercussions in terms of losing advertisers, audience, and in a few cases their jobs.

Chapter 7 discusses the role of critical terrorism coverage and how it often conflicts with the roles the audience and the government expect the press to fill after a terrorist attack. It also hints at First Amendment issues implicit in a push for a patriotic press. The old adage goes that the first casualty of war is the truth. This may also be true when it comes to a different kind of war—the war on terror.

The final chapter outlines lesson learned for future journalists. The relationship between terrorism and the media is worthy of study because a key goal of mass-mediated terrorist attacks is "acquiring the heightened attention of the general public, the political elite, and the decision-making circles."[11] This heightened attention cannot be achieved without the media, and terrorists recognize and understand this fact.[12] Journalists have recognized that terrorists seek out sensational reporting of their actions and use the media as a tool to achieve their aims. Considering this, the discussion of terrorism and journalism in this chapter focuses on ways to improve the coverage.

A record number of people now also seek information about the world from nontraditional news sources such as comedy news shows, entertainment newsmagazines, and talk shows. The different ways that these media approach coverage of terrorism indicates how young people, for example, may be informed about terrorism. Matthew Baum argues that soft news outlets such as entertainment newsmagazines and talk shows have captured viewers who would not normally follow a foreign crisis.[13] These shows transform political issues into entertainment and thus inform a segment of the population that might not normally be made aware. Baum compared hard and soft news coverage of terrorism, particularly September 11, when terrorism became a regular theme of these programs. Interestingly, terrorism was a common topic on these shows even before 9/11: the television newsmagazine Extra had devoted twenty-five programs to non-9/11 terrorism coverage. Television newsmagazine Inside Edition has featured sixteen separate broadcast segments. These entertainment shows tend to focus on individual impact rather than looking at the broad political context in which terrorism occurs. The coverage tends to be episodic, following major events such as the Atlanta Olympic bombing. They profile suspected terrorists and examine potential weapons.[14]

A study of some of the mistakes journalists have made in reporting on past terror attacks has suggested that media organizations develop a set of guidelines for covering terrorism as a means to reduce sensational and panicky headlines, inflammatory catchwords, and speculations about the terrorist plans and government response.[15] Consequently, chapter 8 looks at some of these suggestions and offers insights about how to better navigate the relationship between the press, government, and audience when dealing with terrorism.

Notes

1   National Commission on Terrorist Attacks, The 9/11 Commission Report: Final Report of the National Commission on Terrorist Attacks upon the United States (New York: W.W. Norton, 2004), 20.

2   Brooke Barnett, The War on Terror and the Wars in Iraq, the Greenwood Library of American War Reporting, 8 vols., ed. David A. Copeland (Westport, Conn.: Greenwood Press, 2005).

3   Pew Center Research, "Terror Coverage Boost News Media's Images," Pew Center Research, 28 November 2001, (22 March 2002).

4   Brigitte Nacos, Mass-Mediated Terrorism: The Central Role of the Media in Terrorism and Counterterrorism (Lanham, Md.: Rowman & Littlefield, 2007).

5   Nacos, 36.

6   U.S. Department of State, Patterns of Global Terrorism (1988), v, cited in Ariel Merari, "Terrorism as a Strategy of Insurgency," in Gérard Chaliand and Arnaud Blin, eds. The History of Terrorism: From Antiquity to Al Qaeda (Berkeley, Calif.: University of California Press, 2007), 12–51. Many other scholars use this same definition. See Louise Richardson, ed., The Roots of Terrorism (New York: Routledge, 2006) and Bruce Hoffman, Inside Terrorism (New York: Columbia University Press, 2006), 34.

7   Abraham Miller, Terrorism, the Media and the Law (New York: Transnational, 1982), 1.

8   Bruno S. Frey and Dominic Rohner, "Blood and Ink! The Common-Interest Game between the Terrorists and the Media," Public Choice, 133, 1–2 (October, 2007): 129–145.

9   James W. Tankard, Jr., Laura Hendrickson, J. Silberman, K. Bliss, and Salma Ghanem, "Media Frames: Approaches to Conceptualization and Measurement" (Paper presented at the annual meeting of the Association for Education in Journalism and Mass Communication, Boston, Mass., 1991).

10 David Murray, Joel Schwartz, and S. Robert Lichter, *It Ain't Necessarily So: How Media Make and Unmake the Scientific Picture of Reality* (Lanham Md.: Rowman & Littlefield, 2001).

11 Nacos, 2007.

12 Alex Schmid and Janny de Graaf, *Violence as Communication: Insurgent Terrorism and the Western News Media* (London: Sage, 1982).

13 Matthew A. Baum, *Soft News Goes to War* (Princeton, N.J.: Princeton University Press, 2003).

14 Baum.

15 Raphael Cohen-Almagor, "Media Coverage of Acts of Terrorism: Troubling Episodes and Suggested Guidelines," *Canadian Journal of Communication*, 30 (2005): 3.

CHAPTER 1

What Is Terrorism?

Ter•ror•ism. Noun. *The unlawful use or threatened use of force or violence by a person or an organized group against people or property with the intention of intimidating or coercing societies or governments, often for ideological or political reasons.*

—American Heritage Dictionary

On December 27, 2007, Benazir Bhutto, the former prime minister of Pakistan, was assassinated. Bhutto, who was running for president as an opposition candidate, was killed two weeks before scheduled January 2008 elections. Although differing accounts exist to how she died, the likely cause was a suicide bomb, which also killed twenty-four others. Just two months earlier, Bhutto survived a suicide bomb assassination attempt that killed 136 people.[1]

Assassination is one of the earliest forms of terrorism, according to written records dating back to the Zealot sect, one of the first groups to practice systematic terror in first-century Palestine. In the year 66, the Zealots assassinated several religious and political figures, often using daggers to cut their victims' throats in the midst of a crowd. Like the assassination of Bhutto and many others, the Zealots wanted to "foment a sense of vulnerability within the population at large, a classic tactic of terrorists today. [They] could act wherever and whenever they wanted. That was their strength."[2]

Even better known and better documented than the Zealots is the terrorism practiced by the group known as the Assassins between the eleventh and thirteenth centuries. As Gerard Chaliand and Arnaud Blin note in their definitive work, *The History of Terrorism, from Antiquity to Al Qaeda,* while the Assassins were known for more than their assassinations, their assassination of Nizam al-Mulk, the Persian grand vizier of the powerful Turkish Seljuq sultans, "was one of the greatest terrorist attacks of all time."[3] Chaliand and Blin write, "its contemporary impact was at least as great as that of the assassination of the Archduke Franz Ferdinand or the attacks of September 11, 2001, in their own eras. ... Nizam al-Mulk was a figure of unrivaled repute in the Muslim world of the eleventh century. His place in history had already been ensured by all he had accomplished in his lifetime. In death, he unwittingly opened one of the decisive chapters in the history of terrorism."[4]

Terrorism is a term used to describe many different things by many different people. Political scientists, historians, psychologists, sociologists, criminologists, military scientists, and terrorism scholars have written extensively on both the importance of and difficulty in defining the term. Dr. Ariel Merari, head of the Center for Political Violence at Tel Aviv University, says, "A major hindrance in the way of achieving a widely accepted definition of political terrorism is the negative emotional connotation of the term. 'Terrorism' has become merely another derogatory word, rather than a descriptor of a specific type of activity. Usually, people use the term as a disapproving label for a whole range of phenomena that they do not like, without bothering to define precisely what constitutes terroristic behavior."[5]

Merari further adds that when the term is synonymous with "deplorable" violent behavior its usefulness is "in propaganda" and not in research.[6] Before one can begin to understand the relationship between media and terrorism, he or she must first have a clear working definition of the term, must try to understand the causes of terrorism, and must better understand how terrorism has evolved since the Zealots and the first century of the common era.

Defining Terrorism

Russian leader Vladimir Lenin, responsible for the Red Terror of 1917–1921, once said, "the purpose of terrorism is to produce terror." Most scholars and most working definitions of terrorism acknowledge the simple idea that terrorists produce (or try to produce) terror and fear among opponents. The word terror originates in Latin from the word *terrere*, which means "to frighten." Its first use in the English language was recorded in 1528, but before that it entered into Western vocabularies during the fourteenth century through the French language and was later given modern political meaning during the French Revolution at the end of the eighteenth century.[7]

During the 1980s, two Dutch researchers at the University of Leiden took a social science approach to determining how to best define terrorism. They collected more than one hundred academic and official definitions of terrorism and analyzed them to determine the main components. They found that the element of violence appeared in 83.5 percent of definitions; political goals appeared in 65 percent of definitions; inflicting fear and terror appeared in 51 percent of definitions; arbitrariness and indiscriminate targeting appeared in

21 percent of the definitions; and the victimization of civilians, noncombatants, neutrals, or outsiders appeared in 17.5 percent of the definitions.[8]

Merari found that three common elements exist in the legal definitions of terrorism of the United States, Germany, and Britain: "(1) the use of violence; (2) political objectives; and (3) the intention of sowing fear in a target population." He adds that most academic definitions of terrorism contain these same three cornerstones. Despite the commonalities, Merari argues that these three cornerstone characteristics do not "suffice to make a useful definition" because they're too broad to be useful. He suggests "the main problem is that they do not provide the ground to distinguish between terrorism and other forms of violent conflict, such as guerilla or even conventional war."[9] Merari writes that if a definition of terrorism is equally applicable to all forms of war, including nuclear war, the term loses any useful meaning because it becomes a synonym for "violent intimidation in a political context and is thus reduced to an unflattering term, describing an ugly aspect of violent conflicts of all sizes and shapes, conducted throughout human history by all kinds of regimes."[10] This is one of the criticisms lobbied against many media in their coverage of terrorism—a lack of definitional precision, and a politicizing of the term (which will be highlighted throughout the book). Merari, for practical purposes, uses the U.S. State Department definition of the term, which holds that terrorism is "premeditated, politically motivated violence perpetrated against noncombatant targets by subnational groups or clandestine state agents, usually intended to influence an audience."[11]

Other scholars define terrorism similarly and agree with Merari that a good definition must not lose its original meaning because the term has "become part of the rhetoric of insults exchanged between political opponents."[12] Scholars who contributed to the International Encyclopedia of Terrorism all define terrorism similarly, as "the selective or indiscriminate use of violence in order to bring about political change by inducing fear."[13]

One of the additional challenges in defining terrorism lies in considering the different disciplines that use the term, as well as the context in which terrorism is studied. One of the most widely used definitions of terrorism in criminal justice and military sciences comes from Brian Jenkins, a noted counterterrorism security specialist, who calls terrorism the "use or threatened use of force designed to bring about political change."[14] Walter Laqueur, a criminal justice scholar at Georgetown, offers a similar definition—"the illegitimate use of force to achieve a political objective by targeting innocent people."[15]

Laqueur acknowledges the simplicity of the definition, but argues that it is useless to move beyond simple definitions because the term terrorism is so controversial. Laqueur's own definition exemplifies the challenge in defining the term without adding moral components that complicate the study of terrorism. Because of their subjectivity, words like "illegitimate" and "innocent" would not be found in a definition crafted by scholars like Merari.

Since she edited *Terrorism, Legitimacy, and Power: The Consequences of Political Violence* in 1983, many scholars have cited Martha Crenshaw's important work on political terrorism. In 1983, Crenshaw noted that terrorism cannot be defined without considering the act itself, the target, and the possibility of success. She has consistently written about the importance of defining terrorism by understanding the context in which terrorism occurs. In *Terrorism in Context* she notes, "There are few neutral terms in politics, because political language affects the perceptions of protagonists and audiences, and such effect acquires a greater urgency in the drama of terrorism. Similarly, the meanings of the terms change to fit a changing context."[16]

Terrorism expert Bruce Hoffman writes that the most compelling reason that terrorism is so difficult to define is that the term's meaning has frequently shifted during the past two hundred years. He also argues that today, partly because of the modern media, "most people have a vague idea or impression of what terrorism is but lack a more precise, concrete and truly explanatory definition of the word. ... [V]irtually any especially abhorrent act of violence perceived as directed against society ... is often labeled 'terrorism'" by the media.[17] As Hoffman observes, the media are relevant to any contemporary discussion of terrorism and how it is defined, so much so that some scholars have crafted new definitions of terrorism that account for the impact of the media on public understanding and awareness, often elevating the press role substantially. Political scientist Brigitte Nacos recently crafted her own definition of a form of terrorism she calls "mass mediated terrorism," which she defines as "political violence against noncombatants or symbolic targets which is designed to communicate a message to a broader audience."[18]

What can one make of these variable definitions? Hoffman suggests that "today there is no one widely accepted or agreed-upon definition for terrorism," and adds that when one looks at U.S. governmental agency definitions of terrorism (e.g., the U.S. State Department, the U.S. Federal Bureau of Investigation, the U.S. Department of Homeland Security, and the U.S. Department of Defense), it's not surprising that their definitional variations reflect the different priorities of each agency.[19] The work of other scholars comes to a similar conclusion—that at a minimum, one must start to clarify what terrorism is by at least defining what terrorism is not.

Classifying Political Violence

As previously noted, terrorism loses its meaning when it becomes a catch-all for every form of political violence. Scholars agree that separating terrorism as a strategy for political change from other forms of political violence is useful. Merari's thorough classification of forms of political violence begins by broadly categorizing violence in four ways: state versus state, state versus citizen, citizen versus state, and citizen versus citizen. Typically, state versus state violence takes the form of conventional war, but can include assassination, commando raids, and so on. In all cases of state versus state violence, "these acts are organized and planned and reflect the capability of a large bureaucracy."[20] On the flip side is citizen versus citizen violence, which most commonly occurs in the form of individual crime that is not typically politically motivated.

Terrorism as a strategy generally falls under the remaining two categories—state versus citizen and citizen versus state. Sometimes these two categories are called top-down and bottom-up. Contemporary terrorism is predominantly bottom-up, but historically top-down terrorism has been far more prevalent and has claimed substantially more victims.[21] In top-down, or state versus citizen, two subcategories exist. One is the legal process a state uses to enforce its laws. This is not considered terrorism. However, the other subcategory, which is defined as the "illegal violence used by a government to terrorize and intimidate, usually with the intention of preventing opposition to a regime," is considered a form of terrorism.[22]

The most famous and extreme historical examples of this form of state versus citizen political violence were recorded in Nazi Germany and in the Stalinist Soviet Union. A contemporary example is seen in some Latin American countries' use of death squads that are usually manned by a government's security forces. An offshoot of state versus citizen violence is state sponsored or state terrorism that broadens the forms of violence a state can perpetrate against citizens to instill fear. Chaliand and Blin write that "until very recently, no one spoke of 'state terrorism.' State terrorism, as it is understood today, applies above all to the support provided by certain governments (Libya or Iran, for instance) to terrorist groups, but it takes many other forms."[23] Those forms include using systematic terror as a tool employed by totalitarian regimes (this is the more traditional idea), and sometimes it is found in the military doctrine of a state's armed forces. Chaliand and Blin consider the doctrine of "strategic bombing" as developed in the West in the 1930s as an example of this, because the doctrine was based "entirely on the terror incited by the mass bombing of civilian populations to compel governments to surrender." Chaliand and Blin suggest that it was this doctrine that resulted in the atomic bombing of Hiroshima and Nagasaki.[24] Some would argue this is just one technique used in conventional warfare and is not a form of state terror. Regardless, many scholars point out that the lines between top-down and bottom-up terrorism are often blurred:

> We have all seen today's terrorist become tomorrow's head of state, with whom governments will have to deal at the diplomatic level. Menachem Begin exemplifies this typical metamorphosis. Western tradition considers violence legitimate only when it is practiced by the state. Such a limited definition takes no account of the terror practiced by those who have no other means of redressing a situation they deem to be oppressive. The legitimacy of a terrorist act lies in the objectives of its agents.[25]

State terrorism may be one of the hardest categories for people in democratic societies to understand properly because of the generally accepted notion that the end justifies the means. In democratic societies (more so than in other forms of government), people generally embrace the cause rather than the mode of action, supporting the just war doctrine that legitimates state violent action if the cause, such as the liberation of an oppressed people, is considered to be good. Chaliand and Blin suggest that there is a "dangerous confusion between the moral interpretation of a political act, and the act itself clouds our understanding of the terrorist phenomenon."[26]

Although terrorism clearly occurs in the top-down, state versus citizen category, it is most often observed today as a citizen versus state or bottom-up phenomenon. When it is organized and its aim is to overthrow the government, citizen violence against the state falls under the category of insurgency. Terrorism is one form of insurgency. Other forms include coups, guerilla war, revolution, and riots. A coup is defined as a planned insurgency that takes place at high-level ranks within the state and usually only involves a few people and relatively small amounts of violence over a very brief period of time. Unlike a coup, which is a strategy, revolution typically involves "a change of the system." But, as Merari notes, "under the Leninist model of revolution, revolution does connote a strategy rather than a social or political outcome–it is characterized as an insurgency from below involving numerous people, the period of preparation is very long, but the direct violent confrontation is expected to be brief."[27] Riots are different from all other forms of political violence because they are unplanned. Rioting can recur, but even in these instances it is not considered a "planned, organized, protracted strategy."[28]

Of the varying forms of insurgency, the one most commonly confused with terrorism is guerilla warfare. One of the most influential terrorism handbooks today, the *Minimanual of the Urban Guerilla*, comes from the writings of Carlos Marighella, the Brazilian guerilla revolutionary killed by police in 1969. In Spanish, guerilla is a diminutive meaning "small war." Guerilla warfare is one of the oldest forms of war, even older than conventional notions. Guerilla wars are diffuse, fought in relatively small formations against stronger enemies, and as a strategy of insurgency it avoids direct, decisive battles. Instead, guerillas prefer long, drawn-out struggles, with the idea in some guerilla doctrines that victory comes by wearing out the enemy. As Marighella wrote more than forty years ago, "the primary task of the urban guerilla is to distract, to wear down, to demoralize the military regime and its repressive forces, and also to attack and destroy the wealth and property of the foreign managers."[29]

Some scholars have suggested that guerilla wars are simply an interim phase that allows insurgents to build up a regular army that might have a chance of winning a conventional war. But, typically guerilla warfare involves the use of "hit-and-run operations" with its principle goal of catching the enemy off guard and preventing it from employing its full might against the guerillas. One of the most important differences between terrorism and guerilla warfare is that unlike terrorism, guerilla warfare as a strategy seeks to establish physical control of a territory. Merari explains that

> Notwithstanding the fact that terrorists try to impose their will on the general population and channel its behavior by sowing fear, this influence has no geographical demarcation lines. Terrorism as a strategy does not rely on "liberated zones" as staging areas for consolidating the struggle and carrying it further. As a strategy, terrorism remains in the domain of psychological influence and lacks the material elements of guerilla warfare.[30]

So, what are the fundamental characteristics of terrorism as a distinct strategy of insurgency? As previously noted, no agreed upon definition for the term terrorism exists, but despite the definitional challenges, scholars and terrorism experts do agree that terrorism as a strategy has common elements.

These common elements still allow for the fact that terrorism is "a complex and multivariate phenomenon" that shows up in different forms with different objectives in different parts of the world.[31]

Terrorists operate in very small units, compared to other forms of insurgency. Their numbers can range from the single assassin or the individual suicide bomber to larger teams (typically five or six, but sometimes more as seen in the September 11 terrorist attacks as well as in some elaborate kidnapping plots) that capture hostages or carry out a hijacking. When engaged in battles, which is less common for terrorists than guerilla warriors, they are usually groups of fewer than ten.

In terms of weapons, terrorists typically use improvised devices, such as homemade bombs, car bombs, or barometric pressure-operated devices designed to explode on airplanes. They also use handguns, grenades, and assault rifles but rarely have access to military weapons or artillery. Because terrorists don't have territorial bases like guerillas, they typically work to blend into civilian populations and usually don't wear uniforms.[32] Terrorist targets are state symbols, political opponents, and the public at large. Richardson observes that another common element is that terrorist groups are always significantly weaker than their opponents and "are prepared deliberately to murder noncombatants in furtherance of their objectives. The adoption of terrorism as a tactic to effect political change is, therefore, a deliberate choice."[33] Most agree that terrorism is a strategy of last resort, based on the fact that terrorist groups are very small and because of this often lack the ability to employ other strategies. Finally, terrorism is not recognized as a legal act, either internationally or domestically. Both conventional war and guerilla war are recognized as internationally legal if conducted under specified rules.

Causes of Terrorism

In *The Roots of Terrorism*, a book that grew out of the 2005 International Summit on Democracy, Terrorism, and Security in Madrid, Spain, scholars from a wide range of disciplines tried to identify the root causes of terrorism. They agreed that no single cause exists and that terrorism is a complex problem that requires sophisticated responses.[34]

Psychologist Jerrold Post noted that at an individual level, psychologists have "long argued that there is no particular terrorist personality and that the notion of terrorists as crazed fanatics is not consistent with the plentiful empirical evidence available." He notes that in the field of psychology, scholars have come to consensus on the idea that "group, organizational, and social psychology—and not individual psychology" offer the best lens through which to study the psychological dynamics of terrorism. Post argues that the importance of "collective identity and the processes of forming and trans- forming collective identities cannot be overstated. This, in turn, emphasizes the sociocultural context, which determines the balance between collective and individual identity. Terrorists subordinate their individual identity to the collective identity so that what serves the group, organization or network is of primary importance."[35]

Many terrorism scholars study the collective identity of a terrorist through the stated goals or motivations of a terrorist group. As Merari has observed, most terrorist groups describe themselves at national liberation movements or people who are fighting against some form of oppression, be it social, reli- gious, imperialist, economic, or some combination of these. Much has been made, by both the press and by academics of the quote, "One man's terrorist is another man's freedom fighter." It's difficult to trace the origin of this quote because it is commonly used without citation, but it likely took root as an idea in the 1940s, when Menachem Begin, the former prime minister of Israel, arrived in Palestine and became one of the leaders of the militant Zion- ist group Irgun Zvai Leumi. In the 1940s the Irgun fought against British rule in Palestine, because the Irgun wanted to create an independent Jewish state.[36] Some called Irgun's revolt against the British terrorism. Others called the Irgun freedom fighters. Begin himself spoke about how he saw the differences be- tween a terrorist and a freedom fighter in a speech that was later reprinted in *International Terrorism: Challenge and Response*. Begin says the difference lies in evalu- ating the aims and methods of the two. He argues that the Irgun was not a terrorist group because it took great pains to not harm civilians. In the famous Irgun bombing of the King David Hotel in Jerusalem in 1946, ninety-one people died. The attack was aimed at the British military, which was using the hotel as a military headquarters. Begin says that the Irgun forewarned the Brit- ish of the attack in order to spare lives. He writes, "we gave that warning for half an hour because this is the difference between a fighter for freedom and a terrorist. A terrorist kills civilians. A fighter for freedom saves lives and fights on at the risk of his own life until liberty wins the day."[37] Others, however, still write about the Irgun as a terrorist group, noting that it fit most defini- tions of terrorism and employed the methods used by terrorists, despite the fact that it made some conscious effort to limit noncombatant casualties. In addition to being called freedom fighters and terrorists, the Irgun also has been called a group of guerillas and revolutionaries.

In 1988, then vice president George H.W. Bush introduced a U.S. Department of Defense document called "Terrorist Group Profiles" by asserting that "The difference between terrorists and freedom fighters is sometimes clouded. Some would say one man's freedom fighter is another man's terrorist. I reject this notion. The philosophical differences are stark and fundamental."[38]

The notion that "one man's terrorist is another man's freedom fighter" lives on in popular discussions of terrorism, even within the media, but it may be a misnomer. For example, in correspondence with Honest Reporting, Joanna Mills, editor of BBC World Update, wrote: "It is the style of the BBC World Service to call no one a terrorist, aware as we are that one man's terrorist is another one's freedom fighter." Merari argues that presenting the two terms together "as mutually exclusive is in general a logical fallacy."[39] He says that the terms "terrorist" and "freedom fighter" describe two different aspects of human behavior because one is a method and the other is a cause:

> The causes of groups that have adopted terrorism as a mode of struggle are as diverse as the interests and aspirations of mankind ... Some terrorist groups undoubtedly fight for self-determination or national liberation. On the other hand, not all national liberation movements resort to terrorism to advance their cause. In other words, some insurgent groups are both terrorists and freedom fighters, some are either one or the other, and some are neither.[40]

Of course, much of the difficulty in trying to sort out the causes and principles behind resorting to terrorism as a strategy to effect political change comes back to the idea that the term "terrorism" is so politically loaded. Using Merari's distinction of method and cause, the method is determined and named based on the perceived legitimacy of the cause. This is also true when one examines much of the media coverage of terrorism because the media often adopt the state, political, and/or the cultural notions of what terrorism is. As noted earlier, one of the challenges in discussing and studying terrorism is that terrorism has become so broadly applied to all forms of political violence that many would argue it must be uncategorically condemned as a strategy for political change.

An example of this kind of absolute condemnation is seen in the writings of Benjamin Netanyahu, decades before he became the prime minister of Israel. In 1979, he introduced the Jerusalem Conference on International Terrorism and, in challenging what he called the "easy moral relativism" of the idea that one man's terrorist is another man's freedom fighter, said it was important for the conference "to establish at the outset the fact that a clear definitional framework exists, regardless of political view. Terrorism—the deliberate and systematic killing of civilians so as to inspire fear—was shown persuasively to be, beyond all nuance and quibble, a moral evil, infecting not only those who commit such crimes, but those who, out of malice, ignorance, or simple refusal to think, countenance them."[41]

Despite Netanyahu's assertion that terrorism is a moral evil, historians and other scholars suggest that a "moralistic approach" to terrorism is documented. The "hero" of the moralistic approach is Russian Ivan Kalyayev, who at the turn of the twentieth century was a member of the "combat organization" of the Social Revolution Party. This group used the assassination of government officials as its primary strategy against the czarist regime. On February 2, 1905, Kalyayev was supposed to assassinate Grand Duke Sergei but he did not follow through because Sergei arrived at the proposed site of the assassination with his two young nephews and his wife. Instead, Kalyayev assassinated the Grand Duke several days later when he was alone so as not to harm his innocent family.[42]

Another way to consider the moral implications of terrorism comes from political philosopher Michael Walzer who writes "in its modern manifestations, terror is the totalitarian form of war and politics. It shatters the war convention and political code. It breaks across moral limits beyond which no further limitation seems possible, for within the categories of civilian and citizen, there isn't any smaller group for which immunity might be claimed ... Terrorists anyway make no such claim; they kill anybody."[43] But the idea that terrorism could be moral or immoral is, according to Merari, meaningless. He acknowledges "terrorists wage war by their own standards, not by those of their enemies," and that

> Terrorists usually dismiss the law altogether, without even pretending to abide by it, whereas states pay tribute to law and norms and breach them only under extreme circumstances; but it should be noted that the relativity of morality has been also expressed in the changing rules of combating terrorism. If laws reflect the prevailing moral standards in a given society, one may find interest in the fact that all states, when faced with the threat of insurgency, have enacted special laws or emergency regulations permitting the security forces to act in manners that would normally be considered immoral.[44]

Provoking states into instituting repressive laws or allowing "immoral" practices by military or state security forces was a goal of people like Carlos Marighella, who knew that when states turned to repressive responses that applied to the public at large and not just to insurgents that public support could shift away from the state. This can occur in states with varying forms of government.

It is this kind of political context within which many political scientists study the root causes of terrorism, exploring, among other things, correlations and sometimes causation between terrorism and the type of political environment in which terrorism takes place. Ignacio Sanchez-Cuenca studied revolutionary movements in twenty-one countries during the 1970s and found that terrorist groups emerged in states that had experienced past political instability, had powerful social movements a decade earlier, and had instituted repressive policies. He argues that to understand the conditions under which terrorism exists, scholars should disaggregate the broader concept of terrorism and study particular types of terrorist groups. This is particularly important in understanding international terrorism, he argues.[45]

From a sociological perspective, some scholars have argued that rapid socioeconomic change is a risk factor that may help terrorism exist in certain places because it produces instability and often times dislocates populations. Interestingly, when religious scholars offer perspectives on the causes of terrorism, they note that socioeconomic and political factors remain primary causes of religious and extremist terrorism, but these causes are obscured by the extreme religious language that some terrorists use. Scholar Mark Juergensmeyer has looked broadly at all religions and their relationships to terrorism and concluded that it is economic and political grievances and not religion that are the initial problems. "These secular concerns are now being expressed through rebellious religious ideologies, which makes them more intractable" he says. "These grievances ... are being articulated in religious terms, and are being seen through religious images, and are being organized by religious leaders through religious institutions. Religion then brings new aspects to the conflict."[46]

Religious scholar Karen Armstrong echoes Juergensmeyer's understanding of the relationship between religion and terrorism in some of her recent writing on the subject:

> We rarely, if ever, called the IRA bombings "Catholic" terrorism because we knew
> enough to realise that this was not essentially a religious campaign. Indeed, like the

Irish republican movement, many fundamentalist movements worldwide are simply new forms of nationalism in a highly unorthodox religious guise. This is obviously the case with Zionist fundamentalism in Israel and the fervently patriotic Christian right in the US.[47]

Often, people talk about religious dimensions of terrorism as a modern phenomenon, but actually terrorism with a religious underpinning dates back centuries and is a recurring historical phenomenon. While religious terrorism, particularly the phenomenon of radical militant Islam, is on the rise, terrorism experts seem to agree that "jihadist terrorism cannot prevail, because unlike Islamism, it has no genuine political vision."[48]

The Importance of History

While terrorism dates back to the first century, contemporary ideas about terrorism are usually traced back to the French Revolution of the eighteenth century. Historians say this is not exactly accurate and suggest that the Zealots and the Assassins had quite a bit in common with modern terrorists, and that "terrorism is not a recent phenomenon."[49]

The earliest terrorists were the first-century Jewish Zealots, also called the sicarii, the Latin plural of the word sicarius, which means dagger. This "murderous sect" helped incite an uprising against the Roman occupation of Judea in 70 C.E. that resulted in the destruction of the second temple in Jerusalem. The Zealots assassinated many important religious and political figures, using daggers to cut their victims' throats in the midst of crowded public places. Like modern terrorists, historians have suggested that the actions of the Zealots was intended to send a message to a wider audience, in this case to Roman imperial officials and pro-Roman Jews.

Another religious terrorist sect considered with the Zealots as one of the earliest documented terrorism groups was the Assassins, who were an Islamic correlate that for two centuries (the eleventh through the thirteenth) made its name through the assassination of Muslim dignitaries. The Assassins were not the first "secret society" to turn to assassination and terror, but they were the best organized and longest-lived "terrorist" group operating in that context, even though they were never able to attain central power.[50] The strategy of the Assassins was indirect, based on threats and persuasion and their skill at successfully assassinating:

The indirect strategy, which made a stunning comeback in the twentieth century after the era of revolutionary wars, relies on other than military means in fighting one's enemy. It functions particularly well in a context in which synergy has been achieved between political ends and strategic resources ... The terror that could be inspired by [the Assassins'] deadly assaults was virtually unlimited, given that it was capable of attacking anyone at any time. Even in Europe, some heads of state who had been involved in the Crusades feared for their lives within the shelter of their own castles in England and France. There is no evidence to suggest that the Assassins ever attempted such attacks, but it is the irrational fear that terrorists inspire, out of all proportion to their true capacities to harm, that constitutes their strength.[51]

Chaliand and Blin note that Christian sects did not use terror to the same effect as the Zealots and the Assassins, but that the Taborites of Bohemia in the fifteenth century is an example of a Christian sect that engaged in terrorist activities. They write that all of the early religious terror sects were messianic movements that believed that the world would be transformed by an "event marking the end of history."[52]

As with all other aspects of the study of terrorism, the history of terrorism involves a multitude of terror groups that have functioned in a variety of political contexts with different aims and motivations—some religious, some social revolutionary, some ethnic, some nationalist. Most contemporary discussions about terrorism ignore the Zealots and the Assassins and the many different forms of terrorism that existed around the world prior to the eighteenth century. It is the French Revolution of the eighteenth century that is the starting point for most contemporary analyses of terrorism because that is where the term terrorism and its modern form emerged. While there is no denying the significance of the French Revolution to modern-day understandings of terrorism, several other moments in history and some specific terror tactics that evolved during the time period between the Assassins and the French Revolution are worth noting. They may also help explain some of the complexities of modern state terrorism, as well as the general difficulty in crafting a single, practical definition of what terrorism is.

While the Zealots and the Assassins utilized assassination of political and religious figures as a basic tactic (and, assassination today is still considered an important weapon in a terrorists strategic arsenal), assassination can exist outside of terrorism. Not all political assassinations are defined as terrorist acts. Much of the reasoning for this evolves from the Greek and Roman philosophical perspectives on tyrannicide. While to many tyrannicide is not linked to terrorism, Chaliand and Blin argue that it is linked by history, noting "the direct and indirect influence that the defenders of tyrannicide have wielded over groups engaged in political assassination has been considerable over the centuries."[53]

Tyrannicide is the killing of a tyrant or a despot. In Greek culture, the person who killed a tyrant was a hero. Aristotle did not associate tyrannicide with simple crime because he believed the person who killed a tyrant was actually justified in his action. According to Chaliand and Blin, the "justification of tyrannicide, supported by the desire to bring the act into conformity with the law and to associate it with certain moral rules is akin to the doctrine of the just war, whereby the use of violence is justified in instances that … may be open to interpretation."[54] The acceptance of tyrannicide was an important element of ancient political culture and of the cultures that subsequently arose in Europe and in the Arab world. The Romans were also "fascinated" by the concept of tyrannicide. Cicero had asserted that while assassination was "the most heinous of all crimes," tyrannicide was "the most noble of actions, delivering humanity as it does from 'the cruelty of a savage beast.'"[55]

Others justified tyrannicide in later centuries—John of Salisbury in the twelfth century, Saint Thomas Aquinas in the thirteenth century, and various Catholic and Protestant philosophers in the fifteenth and sixteenth centuries. By the sixteenth century, Jean-Jacques Rousseau's concept of the popular will entered into political philosophy and tyrannicide shifted from being an individual act by a hero, and instead became an act required by the people themselves. It is popular will that motivated people and gave them the right to rise up against a tyrant. English soldier Edward Saxby, who in the seventeenth century plotted to assassinate Oliver Cromwell, wrote in his pamphlet Killing No Murder that the assassination of a tyrant should be undertaken "in the name of the public honor, security and well-being." He argued that tyrannicide is a duty that people must accept on behalf of humanity, and with God's blessing.[56]

Saxby's thoughts about tyrannicide existed more than a decade after the end of the Thirty Years' War in Europe. The signing of the Peace of Westphalia in 1648 helped usher in political realism in Europe, in which another fundamental principle emerged—the noninterference in the affairs of other states because a state is responsible for its own political management no matter what the nature of the state may be, including tyranny. This way of thinking dominated politics until the late twentieth century. But, during the nineteenth century, tyrannicide continued, becoming more than just a way to eliminate tyrants and free people. According to Chaliand and Blin,

> The execution of the tyrant was symbolic, because it opened the way to a purification of the political system and the chance for a new beginning, with the goal not only of changing the political regime but also of transforming society. This new interpretation of tyrannicide … was to mark the entire nineteenth century … For many revolutionary groups and terrorist organizations [today], tyrannicide is a key element of their philosophy.[57]

In addition to the concept of tyrannicide, the idea of terror in warfare emerged during the centuries leading up to the French Revolution. The Mongols, under Genghis Khan and his successor Tamerlane, used terror as a basic tool and strategy for conquest in the thirteenth and fourteenth centuries. They believed that a military victory was not enough and that they had to eliminate their enemy and crush his will to resist. Tamerlane spared people in cities that did not resist him, but when he encountered resistance, he punished that resistance through the massacre of civilians and by raising pyramids built from thousands of decapitated heads to persuade others not to resist.[58]

The use of terror during war in Western Europe didn't emerge until the Thirty Years' War, which occurred between 1618 and 1648. Most of this war was fought in Germany, but it involved most major European powers. The war started as a religious conflict between Protestants and Catholics but gradually evolved into a general war and also fused with a civil war raging in Germany. Because the Thirty Years' War started as a religious war, it involved civilian populations, and despite the fact that the "just war" precepts established by the Church prohibited it, noncombatants became favored targets. At the outset, the war was only one in a series of religious conflicts, but it eventually became a much larger event, with massive armies from across Europe descending on Germany. "As a result of the combination of uninhibited violence and military mass," Chaliand and Blin say, "civilians found themselves in the thick of fighting of which they were the primary victims."[59] Germany's population dropped by 50–60 percent during the decades of the Thirty Years' War, and after decades of fighting, many weary generals hoped to bring the war to a close by terrorizing the enemy. Many gruesome examples of their terrorizing exist, including the French execution of every surviving soldier following a battle in the town of Chatillon-ur-Saone. After the soldiers were executed, their bodies were hanged from the trees in a nearby forest and were left as a reminder, partly because the French believed it "would make the shock effect even stronger and spread the news faster and farther."[60] Although the end of the Thirty Years' War did not end all warfare in Europe, the Peace of Westphalia did put a total end to wars of religion and the terror campaigns that had accompanied them, until the French Revolution began in 1789.

The Reign of Terror during the French Revolution was a turning point in the history of modern terrorism, particularly as it is related to the state. The period began with the execution of deposed King Louis XVI and the ushering in of the rule of the Revolutionary Government in 1793. It was the narrow period of 1793–1794, when the Committee of Public Safety of the National Convention, led by Maximilien Robespierre,[61] was charged with uncovering and foiling conspiracies that the term "reign of terror" was born. The committee ordered the execution of thousands of people, many believed to be innocent, and officially adopted terror as a revolutionary policy.[62] The terror reached its high point when the government executed anyone it considered an enemy of the people and treated those who joined counterrevolutionary movements as members of a foreign enemy state. Historians estimate that as many as thirty thousand people died as a result of the terror, with roughly two thousand six hundred executed in Paris, nearly seventeen thousand executed across France, and the "colonnes infernales" (infernal columns) killing tens of thousands of people considered instigators of the Vendee Rebellion, which included priests, prisoners of war, women, children, and residents of entire villages associated with the rebellion.[63]

Historians note that the "French Terror served both as the founding act of modern state terror and as the model defining and delineating the strategic use of violence by a state apparatus ... The French Terror prefigured a system to be found in all of the great revolutions, especially the Bolshevik Revolution ... the exploitation of ideological fanaticism, the manipulation of social tensions, and extermination campaigns against rebellious sectors."[64]

The founding act of modern terrorism in the French Revolution led to the birth of contemporary terrorism in Russia in the late nineteenth and early twentieth centuries. The series of economic and social upheavals in Russia that led to the eventual overthrow of the tsarist autocracy and the subsequent socialist Provisional Government also eventually led to the Bolsheviks gaining power and establishing the Soviet Union.

It was also in the nineteenth century that the concept of "propaganda by deed" emerged, a practice that was favored by anarchists, particularly in Russia, Italy, Spain, and France. Contemporarily, "propaganda by deed" is understood as the use of acts of violence to gain publicity and enable a terrorist group to spread the word of its insurrection and expand its base. Merari notes that this doctrine has rarely succeeded because terrorists rarely attract public sympathy and support.[65]

Historically, "propaganda by deed" had the most success (albeit limited) in Spain, where anarchist terrorists used the strategy well into the twentieth century. Historian Olivier Hubac-Occhipinti writes that the Spanish anarchist acts of terrorism, unlike the Russian acts during this time, were designed to link the violence with "the doctrine in whose name it had been committed, thus obliging society to acknowledge the intensity of the rage and sentiments of revolt that had motivated it."[66] The Spanish anarchists also targeted people within a certain social class. "The goal here was to kill anyone who collaborated with the bosses or the state or even merely worked within the system. From that standpoint, all members of the bourgeoisie were enemies deserving of death, even if they were assigned no particular responsibility."[67]

In Russia, "propaganda by deed" grew out of the principle of volunteerism, which held that within everyone is a dormant, revolutionary, popular energy that could be "unleashed by propaganda or by terror."[68] Because of their unsuccessful efforts to stir the masses to revolution through pamphlets and other forms of propaganda, the most radical of the anarchists adopted the name Narodnaya Volya (meaning The People's Will) and began to plot the execution of Czar Alexander II. The Narodnaya Volya became the first terrorist group in Russian history. After multiple failed attempts, they finally killed Alexander II with a bomb in 1881, but the power of the act did not produce the desired result of uprising. Public opinion condemned the killing and after several arrests of key leaders of the group, influence of the anarchists fell under Alexander Ulyanov, brother of Vladimir Lenin.[69] Lenin would eventually gain power and lead the Bolshevik government's institution of the Red Terror, modeled after "The Terror" of the French Revolution. The Bolsheviks used terror between 1917 and 1921 to crush their opponents and control their own people. Their terror took the form of concentration camps, executions without trials and taking hostages.

Conclusion

Political scientist Brigette Nacos has widely applied the "propaganda by deed" concept in her writings about terrorism, in which she also highlights the importance of the media to terrorism as it is practiced today. While her work builds on the history and causes outlined in this chapter, Nacos also considers the implications of modern terrorism somewhat differently by adding media to the mix in a central way. Interest in the relationship between the media and terrorism has exploded since September 11, but prior to this, only a small percentage of the scholarly writing about the history, the definitions or the causes of terrorism includes the media as more than just an aside. Generally speaking, the media emerged as a relevant part of the study of terrorism in the 1960s. Nacos is one of the leading political scientists who has explored the significance of the media to understanding how modern terrorism is practiced. In explaining her creation of the term "mass mediated terrorism" she writes:

> the idea here is that most terrorists calculate the consequences of their deeds, the likelihood of gaining media attention, and most important, the likelihood of winning entrance—through the media—to what I call The Triangle of Political Communication. In mass societies in which direct contact and communication between the governors and the governed are no longer possible, the media provide the lines of communication between public offices and the general public.[70]

Nacos writes that modern terrorists have four "media-dependent" goals that include capturing the attention of audiences both inside and outside their "target" societies, seeking recognition for their motives, seeking "the respect and sympathy of those in whose interest they claim to act," and seeking "quasi-legitimate" status so that they receive the same degree of media attention as legitimate political actors.[71]

As suggested by Hoffman, the media are also critically important to the general public's understanding of what constitutes terrorism. He has suggested that "most people have a vague idea or impression of what terrorism is but lack a more precise, concrete and truly explanatory definition of the word" because they learn about terrorism from the media. Yet, the way the media define and report about terrorism varies widely.[72] Understanding the definitional complexities and the long history of terrorism is important to any analysis of media coverage of terrorism. The next chapter explores the ways in which the media create their own definitions of terrorism as well as how terrorism is generally reported by the news media.

Notes

1   CNN, "Benazir Bhutto Assassinated," http://www.cnn.com/SPECIALS/2007/news/
    benazir.bhutto/index.html, viewed January 15, 2008.

2   Gérard Chaliand and Arnaud Blin, eds., The History of Terrorism: From Antiquity to Al Qaeda
    (Berkeley, Calif.: University of California Press, 2007), 58.

3   Chaliand and Blin, 67.

4   Chaliand and Blin, 66–67.

5   Ariel Merari, "Terrorism as a Strategy of Insurgency," in Gérard Chaliand and Arnaud Blin,
    eds. The History of Terrorism: From Antiquity to Al Qaeda (Berkeley, Calif.: University of California
    Press, 2007), 12–51.

6   Merari, 13.

7   Alex P. Schmid, "The Problems of Defining Terrorism," in the International Encyclopedia of Ter-
    rorism (Chicago: Fitzroy Dearborn, 1997), 11–21.

8   Alex Schmid and Albert Jongman, Political Terrorism: A New Guide to Actors, Authors, Concepts, Data
    Bases, Theories and Literature (New Brunswick, N.J.: Transaction Books, 1988), 5–6. Also cited
    in Merari, 3, and Bruce Hoffman, Inside Terrorism (New York: Columbia University Press,
    2006), 34.

9   Merari, 14–15.

10 Merari, 16.

11 U.S. Department of State, Patterns of Global Terrorism (1988), v., cited in Merari, 16. Many
    other scholars use this same definition. See Louise Richardson (Ed.), The Roots of Terrorism
    (New York: Routledge, 2006) and Hoffman.

12 Schmid, 11.

13 Peter Calvert, "Theories of Insurgency and Terrorism: Introduction," in the International En-
    cyclopedia of Terrorism (Chicago: Fitzroy Dearborn, 1997), 135.

14 Jenkins cited in Jonathan R. White, Terrorism: An Introduction (3rd ed.) (Belmont, Calif:
    Wadsworth, 2002), 8.

15 Walter Laqueur, The Age of Terrorism (Boston: Little, Brown, 1987), 72.

16 See Martha Crenshaw, ed., Terrorism, Legitimacy, and Power: The Consequences of Political Violence
    (Middletown, Conn.: Wesleyan University Press, 1983) and Martha Crenshaw, ed., Terror-
    ism in Context (University Park, Penn.: Pennsylvania State University Press, 1995), 7.

17 Hoffman, 1.

18 Brigitte Nacos, Mass-Mediated Terrorism: The Central Role of the Media in Terrorism and Counterterrorism
    (2nd ed.) (Lanham, Md.: Rowman & Littlefield, 2007), 26.

19 Hoffman, 30.

20 Merari, 17.

21 Chaliand and Blin, 6-7.

22 Merari, 17.

23 Chaliand and Blin, 7.

24 Chaliand and Blin.

25 Chaliand and Blin.

26 Chaliand and Blin.

27 Merari, 21.

28 Merari.

29 Carlos Marighella, *Minimanual of the Urban Guerilla*, http://www.marxists.org/archive/marighella-carlos/1969/06/minimanual-urban-guerrilla/index.htm [Accessed February 6, 2008].

30 Merari, 24–25.

31 Richardson, 2.

32 Merari, 25.

33 Richardson, 2.

34 Jerrold Post, "The Psychological Dynamics of Terrorism," in Louise Richardson, ed. *The Roots of Terrorism* (New York: Routledge, 2006). 17–28.

35 Post, 18. For more on the psychological dimensions of terrorism, see Walter Reich, ed., *Origins of Terrorism: Psychologies, Ideologies, Theologies, States of Mind* (Washington, D.C.: Woodrow Wilson Center Press, 1990).

36 Menachem Begin, *The Revolt: Story of the Irgun* (New York: Henry Schuman, 1951).

37 Menachem Begin, "Freedom Fighters and Terrorists," in Benjamin Netanyahu, ed. *International Terrorism: Challenge and Response* (Edison, N.J.: Transaction Books, 1981), 46.

38 George H.W. Bush, *Terrorist Group Profiles*, quoted in Merari, 27.

39 Merari, 27.

40 Merari.

41 Benjamin Netanyahu, *International Terrorism: Challenge and Response* (Edison, N.J.: Transaction Books, 1981), 1–2.

42 Michael Walzer, *Just and Unjust Wars: A Moral Argument with Historical Illustrations* (New York: Basic Books, 1977), 198–199.

43 Walzer, 203.

44 Merari, 31.

45 Ignacio Sánchez-Cuenca, "The Causes of Revolutionary Terrorism," in Louise Richardson, ed. *The Roots of Terrorism* (New York: Routledge, 2006), 71–82.

46 Richardson, 8.

47 Karen Armstrong, "The Label of Catholic Terror Was Never Used about the IRA," http://www.guardian.co.uk/politics/2005/jul/11/northernireland.july7 [Accessed April 16, 2008].

48 Philippe Migaux, "The Roots of Islamic Radicalism," in Gérard Chaliand and Arnaud Blin, eds. *The History of Terrorism: From Antiquity to Al Qaeda* (Berkeley: University of California Press, 2007), 258.

49 Chaliand and Blin, 77.

50 David Morgan, "The Assassins: A Terror Cult," in the *International Encyclopedia of Terrorism* (Chicago: Fitzroy Dearborn, 1997), 40–41.

51 Chaliand and Blin, 77.

52 Chaliand and Blin, 3.

53 Gérard Chaliand and Arnaud Blin, "Manifestations of Terror through the Ages," in Gérard Chaliand and Arnaud Blin, eds. *The History of Terrorism: From Antiquity to Al Qaeda* (Berkeley: University of California Press, 2007), 79. See also Hoffman, 6–7; and *International Encyclopedia of Terrorism*, 29.

54 Chaliand and Blin, 81; Oscar Jaszi and John D. Lewis, *Against the Tyrant: The Tradition and Theory of Tyrannicide* (Glencoe, Ill.: Free Press, 1957), 3–96.

55 Chaliand and Blin, 81.

56 Chaliand and Blin, 83; Henry Morley, ed., *Famous Pamphlets: Milton's Areopagitica, Killing No Murder, De Foe's Shortest Way with Dissenters, Steele's Crisis, Whatley's Historic Doubts Concerning Napoleon Bonaparte, Copleston's Advice to a Young Reviewer and Morley's Universal Library* 43 (New York: George Routledge and Sons, 1886).

57 Chaliand and Blin, 84.

58 David Morgan, "Mongol Terror," in the *International Encyclopedia of Terrorism* (Chicago: Fitzroy Dearborn, 1997), 42–43.

59 Chaliand and Blin, 89–90.

60 Chaliand and Blin, 90.

61 Robespierre was a disciple of Rousseau.

62 Colin Jones, "Terror in the French Revolution 1789–1815," in the *International Encyclopedia of Terrorism* (Chicago: Fitzroy Dearborn, 1997), 48–51. The political atmosphere in France a few years before the Committee came to power allowed terror to take hold in the four or five years leading up to the Reign of Terror. A climate of fear prevailed in France between 1789 and 1815, with the execution of Robespierre in 1794 serving as an important point of decline in violence and terror. But, terror continued intermittently under subsequent regimes. The emergence of "White Terror" also occurred during this period. White terror was the violence, reprisals, and assassinations committed by enemies of the revolutionary regimes.

63 Chaliand and Blin, 102.

64 Chaliand and Blin.

65 Merari, 41.

66 Olivier Hubac-Occhipinti, "Anarchist Terrorists of the Nineteenth Century," in Gérard Chaliand and Arnaud Blin, eds. *The History of Terrorism: From Antiquity to Al Qaeda* (Berkeley: University of California Press, 2007), 120–121.

67 Hubac-Occhipinti, 121.

68 Thomas G. Otte, "Russian Anarchist Terror," in the *International Encyclopedia of Terrorism* (Chicago: Fitzroy Dearborn, 1997), 56–57.

69 Otte, 57.

70 Nacos, 15.

71 Nacos, 20.

72 Hoffman, 1.

CHAPTER 2
The News Media and Terrorism

"[T]errorism and the media are entwined in an almost inexorable, symbiotic relationship. Terrorism is capable of writing any drama — no matter how horrible–to compel the media's attention ... [It] is the media's stepchild, a stepchild which the media, unfortunately, can neither completely ignore [n]or deny."
—Abraham Miller, political scientist, 1982[1]

The way in which terrorism is communicated by the news media as a political act as well as an act of physical and psychological violence is a subject that today generates much debate academically and within the general public. Contemporary media scholars, journalists, political scientists, and terrorism scholars have written dozens of books on the topic, and all agree that the press is an important vehicle through which a majority of people learn about and come to understand terrorism, its causes, and how governments ultimately respond.[2] What is less understood is the degree to which the media's coverage of terrorism has a measurable effect on its audiences (citizens, governments, political leaders, would-be terrorists, etc.). Most research focuses on the relationship between the press, the terrorists, and the government and uses content analysis to determine how the media have covered terrorism and to highlight how a government responds to terrorism.

Based on her multiple content analyses of media coverage of terrorism, political scientist Brigette Nacos argues that the media have become central to terrorist movements. She writes:

> The argument here is not that the news media in general or journalists in particular sympathize with the perpetrators of political violence, but rather that they are unwitting accomplices of media-savvy terrorists. This is not a new development. However, as the move from news-as-information to news-as-entertainment continues, especially in television, media organizations seem increasingly inclined to exploit terrorism as infotainment for their own imperatives (i.e. ratings and circulation). More than ever before, terrorists and the media are in a quasi-symbiotic relationship.[3]

Although Nacos doesn't apply mass communication theoretical language, her argument falls squarely in the area of media sociology—that is, the exploration of the social structural context of press practices that can include sensitivity to ratings, among many other factors, that can influence the stories journalists choose to report as well as how they report them.[4]

To understand fully the complex relationship between the news media and terrorism, it's important to recognize the vast body of mass media research that explains the complexities of modern journalism and the theories through which most mass communication scholars explore media content and media effects. It is the study of the media's coverage of terrorism (most extensively the September 11 attacks) that has led many researchers to make speculative claims about the effects of such coverage. Less research has been conducted to establish any definitive cause and effect relationships between media coverage of terrorism and its impact. Two mass communication theories, agenda setting and cultivation, do generally support some of the leaps that many researchers have made in recent years. Scholars argue that the sheer increase in the number of terrorism-related stories has heightened public awareness of terrorism as an issue on the public agenda, a phenomenon that can be explained by agenda setting theory. Scholars also suggest that the heightened public awareness and increase in media reporting about terrorism has created in some audiences a stronger fear of terrorism (whether real or imagined), which matches the basic tenets of cultivation theory. These two theories, as well as two media content theories—hierarchy of influences and framing—are discussed in this chapter and applied to the ways that the news media cover terrorism.

Media Content Theories

According to mass communication scholars Pam Shoemaker and Stephen Reese, the study of media content is important because it is the basis for determining media impact and predicting media effects.[5] Studying content also serves as an indicator of other underlying forces in communication content, allowing scholars to learn about the people and organizations that produce the news.[6] According to Denis McQuail, "There is potentially much to be learned about the culture, values and living ideology of a society from the totality of mass communication content."[7] One need only to look at U.S. broadcast news media and compare it to Al Jazeera to see how cultural and ideological differences can impact reporting on a wide range of issues, including terrorism.[8]

Shoemaker and Reese have devised a levels of analysis hierarchical approach to studying influences on media content.[9] Those levels include, from the top macro level, the ideological level, the extra-media level, the organizational level, the routines level, and the individual level. Shoemaker and Reese consider the ideological level the outermost level that subsumes all of the others. This level refers to core assumptions about power and the distribution of power in society. Ideology is defined as a "formal and articulated system of meanings, values and beliefs, of a kind that can be abstracted as a 'world view' or a 'class outlook.'"[10] The next level is the extra-media level, at which researchers explore the impact of a journalist's sources, special interest groups, public relations and advertising, technology, and other factors outside of the media organizations or news content. At the organizational level, researchers consider how the news media organization is structured, how media differ, how authority is exercised within the organization, and what effect it has on the resulting content. Specifically, Shoemaker and Reese say, "organizational analysis seeks to explain variations in content that cannot be attributed to differences in routines and individuals."[11] The routine level refers to the influences of the patterned practices of journalism, such as deadlines and news values, on a journalist's work. Finally, the individual level explores how the views, attitudes, training, and background of an individual journalist influence content.[12] Although each level offers different ways to explore influences on content, the approach is hierarchical in that "what happens at the lower levels (individual and routine) is affected by, even to a large extent determined by, what happens at the higher levels" (ideological, extra-media, organizational).[13]

While the weakest level of influence on content is at the individual level, some conditions can exist to raise the impact of an individual journalist's personal ideology on media content. Research has shown that an individual reporter's own biases and beliefs show up more strongly in breaking news scenarios on television than in other contexts because of the need to report news instantaneously and because often reporters themselves become sources.[14] This is useful for better understanding some U.S. televised media coverage of terrorism, because breaking news coverage of terrorism became more prominent in the U.S. news media after the 1995 Oklahoma City attack.

Nacos and others emphasize the important connection between the immediacy of reporting about terrorism at the extra-media and routine levels. Nacos argues, "No other medium has provided more oxygen to terrorism than television because of its ability to report the news instantly, nonstop, and in visuals and words from any place to all parts of the globe, a facility that has affected the reporting patterns of other media as well."[15]

Hoffman suggests that more powerful than television is the Internet, which not only allows for immediate distribution, but also gives more content control to individuals who are not journalists. Hoffman writes that terrorists are aware of the power the Internet gives them:

> In addition to ubiquity and timeliness, the Internet has other advantages [for terror-
> ists]. It can circumvent government censorship, messages can be sent anonymously
> and also quickly and almost effortlessly, and it is an especially cost-effective means of
> mass communication. It also enables terrorists ... to portray themselves and their ac-
> tions in precisely the light and context they wish–unencumbered by the filter, screen-
> ing, and spin of established media.[16]

Pulitzer Prize–winning journalist Stanley Karnow says that the U.S. government needs to adjust its strategy to combat terrorism by factoring in the freedom the Internet provides. He argues that government must also adjust to the speed with which news today travels. The U.S. government's "trouble now is that they're being defeated today not so much by the American press as by technology. I mean, when you get the people appearing on television live over there and issuing their statements [online] about what's wrong with America and so on, or when you get Al Jazeera getting interviews [with terrorists], how do you stop it?"[17] These kinds of questions focus on extra-media influences that not only impact the government's decision making but also the media's. How much should the media report on terrorist-produced content on the Web?

Henry Schuster, CNN's investigative correspondent who covers terrorism, has noted that the Web content produced by terrorist organizations like al Qaeda are increasingly sophisticated and professional and include editorials, news digests, videos (both older and newer) of Osama bin Laden, slick graphics, even videos of the last wills and testaments of suicide bombers. Schuster says, "They premiered on the Internet, one after the other, and were aimed at recruiting Saudi youth." The power of these sites is partially evidenced by the fact that these terrorist Web sites are "as regularly consulted as they are cited (and publicized) by the mainstream press."[18]

Routines and Objectivity

At the routine level, researchers study the patterned, repeated practices and forms that those who work within the media use to do their jobs. Often, routines are constraints and they form the immediate context both through which and within which media workers do their jobs. Routines can be important to study because they often assign meaning and significance to people and events; they "promote a way of looking at events which fundamentally distorts them."[19] Shoemaker and Reese suggest that "as rational, complex organizations with regular deadlines, the news media cannot cope with the unpredictable and infinite number of occurrences in the everyday world without a system ... Organizations must routinize work in order to control it."[20]

A routine can be as simple as a deadline, but it can also involve the selection of sources, adjusting to the speed required to deliver information in a twenty-four-hour news cycle, or routine questions a journalist might ask. After September 11, one new routine of the news media was considering terrorism as a potential cause of explosions, airport disturbances, airplane crashes, and mysterious illnesses, among other events. Steven Chermak noted that the media became so sensitized to terrorism as a potential cause of violence after September 11 that they even entertained it as a cause of the 2003 Columbia space shuttle explosion:

> The news coverage of this tragedy was also interesting because reporters at least entertained the possibility that the disaster was caused by terrorism. It did not matter that Columbia exploded over 200,000 feet above the earth's surface and the level of security to prevent such an attack prior to the launch was great. Although this speculation was soon replaced with an acknowledgement that the explosion could not have been the work of terrorists, the fact that reporters ... initially concluded that the explosion was the work of terrorists provides a window into public consciousness post-9/11.[21]

Nacos writes that the expansion of terrorism coverage since September 11 has created a new routine within which the media, particularly U.S. media, cover any act of political violence, no matter how small. She says, "by devoting extraordinary broadcast time and column inches to even minor violence and elevating it to the level of a spectacular reality show, the mass media, especially television, play into the hands of terrorists."[22]

Often included in discussions about journalistic routines is the notion of objectivity, which is considered a routine as well as a norm. According to Shoemaker and Reese, "objectivity, although a cornerstone of journalistic ideology, is rooted in practical organizational requirements."[23] One accepted definition of objectivity is fairness and balance in decision making, information seeking, and information presentation.[24] Others define objectivity as a ritual—its primary purpose is to defend the product, or media content, from critics.[25] Objectivity has been a cornerstone of American journalism since the mid- twentieth century, yet despite its longevity as a core journalistic principle, the news media are consistently criticized for a lack of objectivity.

Some of this is seen in the ways that terrorism is conceptualized by the news media at the most basic definitional level. As explored in the first chapter, terrorism is a concept that is difficult to define. The editor of the BBC World Update, in referencing a well-known terrorism-related catch phrase, noted that "it is the style of the BBC World Service to call no one a terrorist, aware as we are that one man's terrorist is another one's freedom fighter." Use of the word terrorism in media coverage sometimes conforms to governmental definitions of terrorism, yet even the governmental definitions can vary widely. In addition to the confusion created by variations in governmental definitions, several press organizations have also struggled with the definitional question. These struggles partially grow out of the press routine of objectivity and fairness that compels news media to create a definition that is ideologically appropriate but also derived independently so organizations can maintain credibility. Sometimes, organizations solve the definitional inconsistency by avoiding use of the term "terrorism" completely.

In the early 1990s, the Irish News published in Belfast, Northern Ireland, decided it would drop the use of the word "terrorist" from its coverage of political violence in Ireland and Britain. The paper called the word "stereotypical" and noted that the incidences of violence that occurred in Northern Ireland were different enough in type that the term "terrorism" did not fit them all. The paper was up-front in its denunciation of what it instead called "paramilitary" violence.[26]

Aside from the Irish News' stand on the issue, the Irish press as a whole "generally characterized the activities of paramilitary groups as 'terrorist,' offering a negative representation of the groups and their methods" during the late 1980s and early 1990s.[27] Tim Cooke writes that the Irish media were aware they were a target of the propaganda put forth by the paramilitary groups and resisted having their news agendas manipulated by these groups. At the same time, the newspapers were also keenly aware of the societal and governmental disapproval of paramilitary or terrorist activity of groups like the Irish Republican Army (IRA). As a response to what it saw as "a pattern of news coverage and condemnation which portrayed them as evil, psychopathic, and often irrational," the IRA established its own media to publish its messages. The *Republican News*, published weekly in Dublin, offered statements and interviews from IRA leaders.

It is worth noting that the political face of the IRA, the Sinn Fein Party, was so successful in gaining acceptance and legitimization by the mainstream Irish and British press in the late 1980s and early 1990s that the British government imposed a broadcasting ban on them between 1988 and 1994. The ban included restrictions on the circumstances in which representatives from Sinn Fein, the IRA, and a couple of other "paramilitary" groups could be heard speaking on British television or radio.[28]

Given the rise of reporting about terrorism and the controversy that can come with the coverage of terrorism, many media outlets—particularly international news organizations—have adopted guidelines similar to the *Irish News* policy, developed more than fifteen years ago. BBC's current guidelines to reporting on terror states:

> We must report acts of terror quickly, accurately, fully and responsibly. Our credibility is undermined by the careless use of words which carry emotional or value judgments. The word "terrorist" itself can be a barrier rather than an aid to understanding. We should try to avoid the term, without attribution. We should let other people characterise while we report the facts as we know them ... We should use words which specifically describe the perpetrator such as "bomber," "attacker," "gunman," "kidnapper," "insurgent," and "militant." Our responsibility is to remain objective and report in ways that enable our audiences to make their own assessments about who is doing what to whom.[29]

Reuters has a similar policy, noting that the organization may refer without attribution to terrorism and counterterrorism in general, but that the news organization does not refer to specific events as terrorism. "Nor do we use the word terrorist without attribution to qualify specific individuals, groups or events. Terrorism and terrorist must be retained when quoting someone in direct speech. When quoting someone in indirect speech, care must be taken with sentence structure to ensure it is entirely clear that they are the source's words and not a Reuters label." [30] The Reuters handbook reinforces the BBC approach of using more specific terms such as bomber, hijacker, attacks, attacker, and so on. The handbook further states that "this is part of a wider and long-standing policy of avoiding the use of emotive terms. Reuters does not label or characterise the subjects of news stories. We aim to report objectively their actions, identity and background. We aim for a dispassionate use of language so that individuals, organisations and governments can make their own judgment on the basis of facts."[31]

Some have criticized both Reuters and the BBC for not using the term "terrorism" or "terrorist" when appropriate. In a backward way, the BBC and Reuters approach attempts to address the concerns raised by people who lament the extension of the term terrorism to apply to all forms of political violence, reducing the word "to an unflattering term, describing an ugly aspect of violent conflicts of all sizes and shapes, conducted throughout human history by all kinds of regimes."[32]

In the United States, most media organizations consistently attach the term terrorist or terrorism to al Qaeda and its activities and attacks (including September 11) but do not consistently apply the term "terrorism" and "terrorist" when reporting on other acts of political violence around the globe. Unlike the BBC and Reuters, some major U.S. news services like the Associated Press don't even include the words "terrorism" or "terrorist" in their stylebooks and reporting guidelines. [33] Some local U.S. news media have developed their own guidelines for reporting on terrorism.

Honestreporting.com, a Web site that five British students launched in 2000 then turned over to the Jerusalem Fund of Aish HaTorah to run in 2001, is aimed at highlighting what it considers media bias in the coverage of Israel. The site is pro-Israel and has led a campaign against media outlets that do not refer to Palestinian militant groups as "terrorists." In 2003, the Web site published responses from two Florida newspapers to questions about the newspapers' use of the word "terrorist" to describe al Qaeda, but not to describe Hamas and Islamic jihad.[34] *St. Petersburg Times* editor of editorials and vice president Philip Gailey offered an answer to how he defines terrorism, and he also subsequently instituted a policy change for the paper. That change included using the term terrorism and terrorist in connection with Hamas and Islamic jihad, as well as editing *New York Times* and *Reuters* wire stories published in the *St. Petersburg Times* to include the use of the terms.[35] In defining terrorism, Gailey said, "acts of terror are committed by terrorists, and the horrific bus attack on Israeli civilians, like the dozens of suicide bombings that preceded it, was an act of cold, indiscriminate terror … I don't think militants set out to deliberately kill children."[36] "Militants" is the term the paper had been using to describe Hamas and Islamic jihad.

In the same article, *Orlando Sentinel* public editor and columnist Manning Pynn refused to acknowledge that it would be appropriate to use the term "terrorist" when talking about Hamas or Islamic jihad, but maintained that it was appropriate to use the term to describe al Qaeda. Pynn explained, "the United States was not at war when it was attacked on 9/11; Israel and the Palestinians have been engaged in armed conflict for decades, and moreover, Palestinians are resisting occupation. The use of the word "terror" in the context of Israel would therefore be judgmental and jeopardize impartial news reporting of an ongoing conflict."[37]

Debate about using the term "terrorist" has also arisen in Iraqi media, but the challenges for journalists in Iraq reaches well beyond language. A former Iraqi journalist says that many of the journalists in Iraq regularly fear for their safety, and they know that when they use terms like "terrorism" to report on al Qaeda or other known terrorist groups they put themselves and their families in danger. He says none of the Iraqi newspapers will label an attack as terrorism because of this fear, and because of deep disagreement about how one would define an act of terrorism.[38]

Since 2003, the Committee to Protect Journalists (CPJ) reports that 127 journalists have died in Iraq, 105 of whom were Iraqi. The CPJ estimates that more than 60 percent of these Iraqi journalists were murdered. Of the 127 journalists killed, 98 died as a result of insurgency (including murder and victims of suicide bombs) and 74 died in the Baghdad province.[39] In 2006, Iraqi journalist Bassam Sebti, who writes for the *Washington Post*, noted that his job was so dangerous that he could not tell friends and neighbors about his work:

> I leave home each day, I peer right and left to be sure no one is tracking me. I follow the same routine when I return 12 hours later. Being a journalist for an Iraqi organization is dangerous enough, but working for a foreign news outlet puts you in double jeopardy. In the eyes of insurgents, I am a "spy," an "infidel," a profiteer exploiting the suffering of Iraqis.[40]

When Westerners discuss Arab media, most immediately think of Al Jazeera, the Qatar-owned Arabic satellite TV news network that became a household name soon after the September 11 attacks for its exclusive broadcasts of the first post-9/11 video footage of Osama bin Laden and a few months later with exclusive footage of U.S. air strikes against Afghanistan.[41] Al Jazeera emerged in 1996, and from the beginning it was received by the United States and moderate Arab states as a "mouthpiece" for the Taliban and al Qaeda, two groups that most U.S. media label as terrorist. According to scholars who have written extensively about the controversial network,

> Arab views toward Al-Jazeera vary tremendously, with some Arab regimes accusing it of being an avenue for dissident voices and a conspirator in antigovernment movements, others acknowledging the network as the sole voice of journalistic objectivity in a conflict-ridden region ... As the so-called war on terrorism makes headlines worldwide, the hot story in political and media circles is not only the war on the ground; it is also Al-Jazeera, the first twenty-four hour all-news network in the Arab world.[42]

Al Jazeera provides extensive coverage of terrorism, but like Reuters, the BBC, and Iraqi media, the network doesn't use the term "terrorism" in its coverage. Al Jazeera's policy on reporting terrorism is not publicly available, but abundant evidence of the television station's policy not to call events "terrorism" or groups "terrorists" exists in its daily reporting. Stories refer to al Qaeda "fighters" and to "bombings," "suicide bombings," "attacks," and other forms of violence, but use of the words "terrorism" and "terrorist" are notably absent.[43] Al Jazeera's caution in using the terrorism label could come from BBC influence—the network uses the BBC as a prime model and many of the Al Jazeera correspondents came to the network from the BBC.[44] Or, it could come from Al Jazeera's experiences in covering acts that some would call terrorism. Historically, the network has been accused of a strong pro-Palestinian bias because it generally supports the Arab position that the Israeli "occupation" of Palestine is illegal. In its coverage of the ongoing violence between Israel and Palestine, Al Jazeera has aired multiple interviews with Israeli leaders but has also chosen not to use the term terrorist to describe Palestinian suicide bombers. Instead, the network calls them "shuhada" (martyrs), a term that has created some controversy and generated great criticism against the network.[45]

Al Jazeera's perspective on terminology might also be informed by important Arab leaders. In an October 24, 2001, interview with Libyan leader Muammar Qaddafi, Al Jazeera broadcast Qaddafi saying that while the United States had every right to retaliate for the September 11 attacks, he would not label Osama bin Laden a terrorist until "an international conference agreed on a definition of 'terrorism.'" He added, "We must sit down at any level without emotions ... and after we define terrorism we agree on fighting terror-ism."[46]

No matter what the source of Al Jazeera's move away from using the term "terrorism," it provides another example of the importance of language and an organization's role in crafting how news about terrorism is framed. All of the definitional challenges noted here reflect organizational and routine-level issues that impact the resulting news content.

While most discussions about reporting policies and defining terms occur at the organizational and routine levels, sometimes definitional challenges emerge at the individual level. Individual levels of influence are considered factors intrinsic to a journalist. They include personal and professional back-grounds; education; personal attitudes, values, and beliefs; and professional orientations, role conceptions, and ethics.[47] Roles are norms that apply to specific types of behavior that are expected of people occupying certain posi-tions in society.[48] Media scholars David Weaver and G. Cleveland Wilhoit have defined three basic journalistic role conceptions—interpretive, dissemi-nation, and adversary—in their surveys of American journalists.[49] Reporters who believe their function is interpretive focus on analyzing complex prob-lems, discussing national policy, and investigating official claims. Reporters who believe their role is to disseminate information say that getting informa-tion to the public quickly is the most important function. Finally, reporters who are adversarial believe their role is to serve as an adversary to business or government officials. Because of the overlap among the roles, most modern journalists "see their professional role as highly pluralistic."[50] In the 1996 American Journalists survey, Weaver and Wilhoit expanded the idea behind the "disseminator" role, saying that "with technical requirements of disseminat-ing great volumes of descriptive information ... the disseminator function is even more focused on immediacy and speed than a decade ago and is a very strong streak in more than half of contemporary journalists."[51]

Shoemaker and Reese observe that when communication workers have more power over their messages and work under fewer constraints "their per-sonal attitudes, values and beliefs have more opportunity to influence con-tent."[52] At least a decade before breaking news became a staple of television news, Herbert Gans noted that although journalists try to be objective, "nei-ther they nor anyone else can in the end proceed without values. Further-more, reality judgments are never altogether divorced from values."[53] He suggests that values and attitudes are mostly kept in check by organizational routines. As seen in the preceding discussion of terrorism definitions, organizations establish when and how the term is used; individual reporters do not make judgment calls.

But, in the first six hours of television news coverage of the September 11 attacks, one could see how some journalists were making quick value judgments about what had happened. Here's an example, from then ABC news anchor Peter Jennings: "As you look at these scenes, you can feel absolutely clear that you are looking at the results of the United States at war with angry and vicious people who will do in the future as they have in the past, whatever they can to get at the United States, this huge presence in the world."[54] The ability of individual journalists to assert their judgments about what was happening that day showed that changes in the routine and organizational expectations of offering news and analysis "first" can lead to more powerful influences on content at the individual level.

The theoretical perspective known as social constructionism or the social construction of reality also helps explain the reason a journalist would naturally draw heavily on his or her professional experience, personal values, attitudes, and ideologies absent an established routine. This theory suggests that people create reality, and the world that they believe exists is based on their individual knowledge and from social interactions with other people.[55] Within such a view a journalist would quite naturally draw on professional and personal values to shape a given news story. This is also true when one considers how the audience perceives news.

When defending a study he coauthored about journalists and their personal religious orientations, Robert Lichter wrote, "news judgment is subjective and decisions about sources, news pegs, and ... language will partly reflect the way a journalist perceives and understands the social world."[56] This process was clearly demonstrated in CNN's "live" coverage of the Oklahoma City bombing in 1995 and in the way all of the networks handled the initial reporting of the events of September 11.[57]

An example of an individual journalists' judgment trumping an organization's came soon after the September 11 attacks when then Fox News Channel (FNC) reporter and anchor Geraldo Rivera was interviewed by journalists on CNBC and FNC. Rivera said that he was a "changed man" by the September 11 terrorist attacks. He defended his view that the pursuit of former President Bill Clinton for his affair with an intern in the late 1990s was illegitimate by suggesting it was partly culpable for the terrorist attacks. Rivera was quoted as saying, "I would bet you that I can find you 4,000 [or] 5,000 FBI agents who wish to God they weren't assigned to Whitewater, Monicagate, Bill Clinton—that instead they were on the trail of Osama bin Laden and the people who were plotting mass murder against us."[58] Rivera's comments triggered responses from other journalists, many of whom believe in the routine of objectivity, and suggested that such comments by a journalist covering the "war on terror" were inappropriate.

Framing

The theory of framing in mass communication is still evolving. Framing is the one theoretical concept that many political science and mass communication scholars have repeatedly applied to the study of media coverage of terrorism. Framing is defined as "a central organizing idea for news content that supplies a context and suggests what the issue is through the use of selection, emphasis, exclusion, and elaboration."[59] According to Robert Entman, "to frame is to select some aspects of a perceived reality and make them more salient in a communicating text, in such a way as to promote a particular problem, definition, causal interpretation, moral evaluation, and/or treatment recommendation for the item described."[60] In thinking about the impact of news framing, media scholars have noted that "very powerful concepts, central to frames, need not be repeated often to have great impact."[61]

In *Framing Terrorism*, several scholars quantitatively and qualitatively explore the different ways that news media have framed news stories about terrorism. Overall, the book theorizes that "the events of 9/11 can best be understood as symbolizing a critical cultural shift in the predominant *news frame* used by the American mass media for understanding issues of national security, altering *perceptions* of risks at home and threats abroad." The authors add, "what changed, and changed decisively with 9/11, were American perceptions of the threat of world terrorism more than the actual reality."[62]

Some of this shift in the predominant news frame is seen in the routine changes of journalism noted earlier in this chapter. Terrorism became a part of the daily news cycle in the United States after September 11, and it entered into some of the routine questions journalists asked at disaster and other crime scenes. In addition, the news media after September 11 adopted the "war on terrorism" frame that was put forth by the Bush administration. The "war on terrorism" frame differentiated countries as "friends" and "enemies" of the United States and was used to justify the U.S. wars in Afghanistan and Iraq. As

President Bush put it, "Every nation in every region now has a decision to make. Either you are with us, or you are with the terrorists."[63]

Some of the media framing studies have explored differences in international press coverage of terrorist attacks. One such study compared U.S. and African newspaper coverage of September 11 as well as the 1998 bombings of the U.S. embassy in Nairobi, Kenya, and Tanzania. The research showed that the geographic closeness of the news organization to the terrorist attack impacted the prominence and amount of coverage and that cultural influences played only a modest role in the reporting of these terrorist events. The study concluded that "national worldviews derived from the international system greatly shape the interpretation and framing of the causes of the terrorist attacks."[64] Similarly, in a comparative framing analysis of CNN and Al Jazeera's coverage of the war on terrorism in Afghanistan, meaningful differences emerged. "CNN stories followed the format of typical international crisis stories, similar to the Persian Gulf War, with frames of consensus, a focus on strategy, technological precision, and a euphemistic description of events."[65] Al Jazeera offered coverage from a different cultural perspective but "also rallied its viewers by calling for unification of the Arab world in international issues. Further, it mimicked some of the same frames of American technological superiority. But the primary contrast in news coverage and framing is that Al-Jazeera did not gloss over a humanistic portrayal of the consequences of war."[66]

The use or lack of use of the term "terrorist" and "terrorism" can also be considered a question of framing. Nacos writes that the U.S. media are more prone to call an act of political violence terrorism if a U.S. citizen is a victim. She reports that a content analysis of three leading U.S. newsmagazines between March 1980 and March 1988 showed that the "terrorism" label was used in 79 percent of the cases that involved American victims, but only in 51 percent of cases that did not include an American victim. She also argues that the U.S. media are inconsistent in the way they frame domestic versus international terrorism. Nacos references the murder of abortion doctor Barnett Slepian in 1998 as an example. Antiabortion extremists killed Slepian because he performed abortions. Nearly all U.S. media who reported on the case considered the murder a crime, but not an act that rose to the level of terrorism. The word terrorism was not used in domestic news coverage of the Slepian killing.[67]

Immediately following the September 11 terrorist attacks in New York and Washington, DC, the BBC World Service adhered to its policy not to use the term terrorism in its coverage and received criticism from U.S. media but support from international journalists for the decision. At the heart of the debate were questions about framing, and U.S. news managers saw the answers differently than international news managers. At a Newsworld Conference debate in Barcelona about the television coverage of September 11, the BBC World Service's deputy director of news, Mark Damazer, asserted that "However appalling and disgusting [September 11] was, there will nevertheless be a constituency of your listeners who don't regard it as terrorism. Describing it as such could downgrade your status as an impartial and independent broadcaster."[68] He added that the BBC World Service had a strong reputation for impartiality and that it "has to be careful about its use of language. It does not usually describe IRA attacks as terrorism, because they may not be seen as such in a world context."[69]

In response, Tony Burman, the executive director of the Canadian Broadcasting Corporation (CBC), agreed and further criticized American news media for the way they were covering the "War on Terrorism" in Afghanistan. He said, "U.S. coverage of the crisis had failed to take account of the international perspective: It's depressing to see the jingoism, which is lamentably part of the culture and spirit of the coverage."[70]

NBC News' vice president Bill Wheatley rejected that charge and said, "It's true that U.S. networks are focusing on the attempt to defeat the Taliban and apprehend Osama bin Laden, but I don't think we've been pulling punches in terms of the difficulties of the war effort and the problems of U.S. foreign policy."[71] He then criticized Al Jazeera's coverage as impartial because of the access the network enjoyed in Afghanistan. "They have been given special status in Kabul, [and] we feel it's correct that our viewers know that they have that special access." Burman, the CBC executive, said that watching the BBC coverage of the war in Afghanistan and comparing it to U.S. coverage was like watching "two different wars."[72] The debate between the CBC, the BBC, and the NBC executives about coverage of terrorism by BBC, NBC, and Al Jazeera is a good example of framing at work as Entman defined it earlier: "[T]o frame is to select some aspects of a perceived reality and make them more salient in a communicating text, in such a way as to promote a particular problem, definition, causal interpretation, moral evaluation, and/or treatment recommendation for the item described."[73]

Agenda Setting

Some scholars discuss framing as the second level of agenda setting. Agenda setting theory holds that the issues (often called "objects") emphasized in the news come to be regarded over time as important by members of the public.[74] Walter Lippmann's opening chapter in his 1922 classic Public Opinion— "The World Outside and the Pictures in Our Heads"—summarizes the agenda setting idea even though he did not use that phrase. His thesis was that the news media, our windows to the vast world beyond our direct experience, determine our cognitive maps of that world. Lippmann argued that public opinion responds not to the environment, but to the pseudo-environment, the world constructed by the news media.[75]

Initially, agenda setting studies involved comparing content analyses of news with public opinion polling data to match the media and the public's agendas during political elections. This is now known as first-level agenda setting, because the focus is on determining the salience of objects and not the attributes of an object. But, any object on the media or public agenda has attributes, characteristics, and properties that describe it. Just as issues on the agenda vary in importance, so do their attributes. One important part of the news agenda is the attributes that journalists and, subsequently, members of the public, have in mind when they think about and talk about an issue. These attributes have two dimensions, a cognitive component regarding information about substantive characteristics that describe the object and an affective component regarding the positive, negative, or neutral tone of these characteristics on the media agenda or the public agenda. The influence of the attribute agendas in the news on the public's attribute agenda is called the second level of agenda setting.[76]

Since mass communication scholars Maxwell McCombs and Donald Shaw articulated agenda setting theory nearly 40 years ago, researchers have conducted more than 425 empirical studies on the agenda setting influence of the news media. This vast accumulated evidence comes from many different geographic and historical settings worldwide and covers numerous types of news media and a wide variety of public issues.[77] It is no longer limited to the exploration of news agendas during elections. Many studies in recent years have applied agenda setting theory to the study of how the news reports on terrorism. These first-level studies have concluded that when the news media cover terrorism heavily and prominently, the public ranks terrorism high on its list of national problems. But, when the coverage declines and is less prominently displayed, the public ranks terrorism lower as a national problem.[78]

A 2003 study that explored news coverage of terrorism in the *New York Times*, ABC, CBS, NBC, and CNN one year before and one year after September 11 and compared it to the Gallup Report's "what is the most important problem facing the country" question during the same time period found that "public concern mirrored the network news coverage."[79] The number of respondents who answered "terrorism" to "what is the most important problem facing the country" shot up from zero in the three months before September 11, 2001, to 46 percent immediately following September 11, 2001. The study cannot conclusively say that the media coverage of September 11, and not the event itself, caused this change, but the similarities in media coverage to public opinion did demonstrate a strong first-level agenda setting effect.

Agenda setting theory acknowledges that the news media are not the only source of information or orientation to issues of public concern. Within agenda setting theory, issues are divided between those that are obtrusive and unobtrusive. Obtrusive issues are issues people personally experience. Unobtrusive issues are those that we encounter only in the media and not in our daily lives. Terrorism, for most people, is an unobtrusive issue. Some issues can be both obtrusive and unobtrusive. Generally speaking, the strength of agenda setting effects of the mass media are strongest when dealing with unobtrusive issues, because the media are often the sole source of that information.[80]

David Weaver developed the concept of need for orientation based on psychologist Edward Tolman's theory of cognitive mapping to offer a richer psychological explanation for variability in agenda setting effects than simply classifying issues along the obtrusive/unobtrusive continuum. Tolman suggests that we form maps in our minds to help us navigate our external environment. The need for orientation concept holds that there are individual differences in the need for orienting cues to an issue and in the need for background information about an issue. Conceptually, an individual's need for orientation is defined in terms of the perceived relevance of an issue and the level of uncertainty toward an issue.[81]

Relevance is considered the initial defining condition. In situations when the relevance of an issue to an individual is low, the need for orientation is low. Among individuals who perceive a topic to be highly relevant, the level of uncertainty must also be considered. If a person already possesses all the information he or she needs about an issue, uncertainty is low. Under conditions of high relevance and low uncertainty, the need for orientation is moderate. When relevance and uncertainty are high, however, need for orientation is high. This is often the situation during primary elections, when many unfamiliar candidates clutter the political landscape. As one might guess, the greater a person's need for orientation, the more likely he or she will attend to the news media agenda, and the more likely he or she is to reflect the salience of the objects and attributes on the news media agenda.

The application of need for orientation to the impact of media coverage of terrorism is seen in a study that Tamar Liebes and Anat First undertook that explores the framing of coverage of the Palestinian-Israeli conflict. They examined the power of the image of Muhammad Dura, a Palestinian boy who was killed in his father's arms in 2000, supposedly by Israeli fire. The image was captured by a Palestinian videographer and broadcast on French television. It was aired on U.S. television weeks later. The media used the image of Dura to symbolize the human cost of the ongoing Israeli-Palestinian struggle. The image was used to put a human face on the ongoing violence between Israel and Palestine, but it was "removed from the relevant sequence of events, as well as [its] political and historical context." Ultimately, it was discovered that Dura died as a result of Palestinian fire, and Liebes and First note that "the understanding that these pictures deliver is particularly useful to viewers who are unfamiliar with the context from which they have been extracted."[82] In a need for orientation model, the potential impact of images like that of Dura is high given that many people in the United States consider the ongoing Israel-Palestine conflict highly relevant but also personally unfamiliar. Few terrorism studies address the need for orientation concept directly, but it is a useful concept through which to learn how much potential impact news coverage may have on audiences. If the statistics are accurate and few people have firsthand experience with or exposure to terrorist activity, yet media coverage is intense, it may be reasonable for scholars like Nacos to argue that powerful media effects on the audience occur.

Cultivation

Cultivation theory holds that television, as a central source of information and culture in America, teaches its audience a "common worldview, common roles, and common values."[83] George Gerbner and his colleagues at the Annenberg School of Communication at the University of Pennsylvania developed cultivation theory in the 1970s. They based their research on comparisons of heavy and light television viewers. Gerbner's research has shown that differences in how heavy and light television viewers see the world do show up across a number of other important variables, including age, education, reading the news, and gender. Still, some critics at the time suggested that Gerbner had not done a sufficient job of controlling for these variables in his research,[84] which led Gerbner to revise the theory in 1980 and add the concepts of "mainstreaming" and "resonance."[85] Gerbner suggested that "mainstreaming" occurred when heavy television viewing led to a convergence of views across groups; "resonance" occurred when the cultivation effect was boosted in certain groups. For example, Gerbner and his colleagues found that both high- and low-income groups who were heavy viewers of television considered fear of crime a serious personal issue—their views were "mainstreamed." Yet, the two groups did not share "mainstream" views about fear of crime when they were light television viewers. Gerbner discovered "resonance" when comparing men and women and their fear of crime. Women who were heavy viewers of television exhibited the strongest responses to the idea that fear of crime was a serious problem, likely because of their perceived vulnerability to crime "resonated" with the television version of the world as high crime.[86]

The addition of mainstreaming and resonance did substantially modify the theory, and today cultivation theory does not claim "uniform, across-the-board effects of television on all heavy viewers. It now claims that television interacts with other variables in ways such that television viewing will have strong effects on some subgroups … and not on others."[87] Most scholars still apply cultivation theory to the study of fear of crime and some have used cultivation to explore fear of terrorism as a result of increased media coverage.

Some media scholars, when they discuss cultivation theory, also mention Marshall McLuhan's notion of media determinism as separate but related. McLuhan became famous in the 1960s for his phrase, "the medium is the message."[88] McLuhan was interested in how media technology altered people's patterns of perception. McLuhan wrote much about television as a tactile medium that involved the senses and was more participatory than print media. Ultimately, McLuhan suggested that the most important effects of a medium came from form and not content, and McLuhan theorized that television would over time have the most pervasive and far-reaching effects.[89] Some scholars dismissed McLuhan's ideas because he basically said content didn't matter. But, today, more scholars are exploring both form and content to determine how media effects might differ by medium.[90] Chapter 5 highlights the special role that television plays in our understanding of terrorism.

The general ways that news media coverage of terrorism may impact audiences' fear of terrorism via cultivation theory is something that scholars have suggested but it hasn't been extensively studied. In Framing Terrorism, scholars compared real world data about terrorism incidences with news media reports about terrorism. The findings suggest that the U.S. government exaggerated claims that terrorism worsened after September 11 and that the George W. Bush administration's suggestion that "Americans live in an especially dangerous place" was statistically untrue based on the U.S. State Department's own data.[91] The public misperceived the statistical risk of terrorist acts, based on government assertions and media coverage, most notably television coverage. The researchers wrote, "the power of consensual news frames, exemplified by the 'war on terrorism' frame in America cannot be underestimated." They added, "A one-sided news frame can block the reception of contrary independent evidence … The events [of September 11], depicted so vividly live on American television screens, carried tremendous visual immediacy and symbolic weight."[92]

Conclusion

Two common terms used to describe the relationship between the media and terrorism are "symbiotic" and "paradoxical." Many scholars are critical of the attention the media pays to terrorist activities and attacks. Hoffman writes, "The modern news media, as the principal conduit of information about such acts, thus play a vital part in the terrorists' calculus. Indeed, without the media's coverage the act's impact is arguably wasted, remaining narrowly confined to the immediate victim(s) of the attack rather than reaching the wider 'target audience' at whom the terrorists' violence is actually aimed. Only by spreading the terror and outrage to a much larger audience can the terrorists gain the maximum potential leverage that they need to effect fundamental political change."[93]

While nearly everyone agrees that the communication of the terrorist act is an important component to the terrorist strategy (dating back to the emergence of the "propaganda by deed" concept during the late nineteenth century), research about the impact of media coverage of terrorism is far from conclusive about its effects. As some of the theoretical concepts in this chapter show, identifying media content is only one step in moving toward a better understanding of media effects. As Shoemaker and Reese observe, "research and theory in mass communication have focused on media effects or, even more often, the effects of media use. ... We suggest ... linking influences on content with the effects of content [to] help build theory and improve our understanding of the mass communication process."[94] This also holds true for the study of news media and terrorism. Without empirically linking content to effects, researchers can't do much more than speculate.

Some scholars have noted that governments are equally adept in framing the way that media covers terrorism. The use of the "war on terrorism" frame in American media coverage was generated by the White House and resulted in tremendous public support for governmental policies aimed at combating terrorism.[95]

One of the questions that is often asked in mass media research is "who sets the media's agenda?" While some would argue the terrorists are doing so because of the media's thirst for drama and ratings ("Terrorism is theater ... carefully choreographed to attract the attention of the electronic media and the international press," wrote Brian Jenkins in 1975),[96] others would counter that the government exerts stronger influence on what the media reports. The next chapter will explore the relationship between the media and government.

Notes

1   Abraham Miller, *Terrorism, the Media and the Law* (New York: Transnational, 1982), 1.

2   Too many books about terrorism and the press exist to list them all here, but it is worth noting some of the more significant works. These include Yonah Alexander and Robert Picard, *In the Camera's Eye: News Coverage of Terrorist Events* (Washington, D.C.: Brassey's, 1991); Bethami A. Dobkin, *Tales of Terror: Television News and the Construction of the Terrorist Threat* (New York: Praeger, 1992); Richard Schaffert, *Media Coverage and Political Terrorists: A Quantitative Analysis* (New York: Praeger, 1992); David L. Paletz and Alex P. Schmid, eds., *Terrorism and the Media* (Newbury Park, Calif.: Sage, 1992); Robert G. Picard, *Media Portrayals of Terrorism: Functions and Meaning of News Coverage* (Iowa: Iowa State University Press, 1993); Gabrielle Weimann and Conrad Winn, *The Theater of Terror: The Mass Media and International Terrorism* (New York: Longman/Addison-Wesley, 1994); Brigette L. Nacos, *Terrorism & the Media: From the Iranian Hostage Crisis to the Oklahoma City Bombing* (New York: Columbia University Press, 1996); Bradley S. Greenberg and Marcia Thomson, *Communication and Terrorism: Public and Media Responses to 9/11* (Cresskill, New Jersey: Hampton Press, 2002); Stephen Hess and Marvin Kalb, eds., *The Media and the War on Terrorism* (Washington, D.C.: Brookings Institution Press, 2003); Pippa Norris, Montague Kern, and Marion Just, eds., *Framing Terrorism: The News Media, the Government, and the Public* (New York: Routledge, 2003); Nancy Palmer, ed., *Terrorism, War and the Press* (Cam-

bridge, Mass.: Joan Shorenstein Center on the Press, Politics and Public Policy, 2003);
Danny Schechter, *Media Wars: News at a Time of Terror* (Lanham, Md.: Rowman & Littlefield,
2003); Nacos.

3 Nacos, 38.

4 Pam Shoemaker and Stephen D. Reese, *Mediating the Message: Theories of Influence on Mass Media
Content* (New York: Longman, 1996); Stephen D. Reese, "Understanding the Global Journal-
ist: A Hierarchy-of-Influences Approach," *Journalism Studies*, 2 (2001): 173–187. Some clas-
sic media sociology studies include Warren Breed, "Social Control in the Newsroom: A
Functional Analysis," *Social Forces*, 33 (1955): 326–335; David Altheide, *Creating Reality: How
TV News Distorts Events* (London: Sage, 1976); Herbert J. Gans, *Deciding What's News* (New York:
Pantheon, 1979); John W.C. Johnstone, Edward J. Slawski, and William W. Bowman, "The
Professional Values of American Newsmen," *Public Opinion Quarterly*, 36 (Winter 1972–1973):
522–540; Mark Fishman, *Manufacturing the News* (Austin, Tex.: University of Texas Press,
1980).

5 Shoemaker and Reese.

6 Denis McQuail, *Mass Communication Theory: An Introduction* (London: Sage, 1994).

7 Denis McQuail, *Sociology of Mass Communications* (Middlesex: Penguin Books, 1972), 14.

8 For more information about and the history of Al Jazeera, see Mohammed El-Nawawy and
Adel Iskandar, *Al-Jazeera: The Story of the Network That Is Rattling Governments and Redefining Modern Journalism* (Cambridge, Mass.: Westview Press, 2003).

9 Shoemaker and Reese.

10 Raymond Williams, *Marxism and Literature* (New York: Oxford University Press, 1977), 109.

11 Shoemaker and Reese, 139.

12 Shoemaker and Reese.

13 Shoemaker and Reese, 12.

14 Amy Reynolds and Brooke Barnett, "'America under Attack' CNN's Visual and Verbal Framing of September 11," in Steven Chermak, Frank Bailey, and Michelle Brown, eds. *Media Representations of September 11th* (New York: Praeger, 2003), 85–101; Amy Reynolds and Brooke
Barnett, "This Just In ... How National TV News Handled the Breaking Live Coverage of
September 11th," *Journalism & Mass Communication Quarterly*, 80 (2003): 689–703.

15 Nacos, 47.

16 Bruce Hoffman, *Inside Terrorism* (New York: Columbia University Press, 206), 201–202.

17 Hess and Kalb, 24.

18 Hoffman, 226.

19 Altheide, 24.

20 Shoemaker and Reese, 118.

21 Steven Chermak, "The Presentation of Terrorism in the News" (Paper presented at the International Conference on the TV Presentation of Crime in Milan, Italy, May 15–16, 2003),
1.

22 Nacos, 222.

23 Shoemaker and Reese, 112.

24 Dan Drew, "Roles and Decisions of Three Television Beat Reporters," *Journal of Broadcasting*, 16
(1972): 165–173.

25 Gaye Tuchman, "Objectivity as Strategic Ritual: An Examination of Newsmen's Notions of Objectivity," *American Journal of Sociology*, 77 (1977): 660–679.

26 Tim Cooke, "Paramilitaries and the Press in Northern Ireland," in Pippa Norris, Montague Kern, and Marion Just, eds. *Framing Terrorism: The News Media, the Government, and the Public* (New York: Routledge, 2003), 75–90.

27 Cooke, 78.

28 Cooke, 85.

29 BBC Editorial Guidelines, "War, Terror & Emergencies," available online at http://www.bbc.co.uk/guidelines/editorialguidelines/edguide/war/mandatoryreferr.shtm l [Accessed March 14, 2008].

30 Sean Maguire, "When Does Reuters Use the Word Terrorist or Terrorism?" available online at http://blogs.reuters.com/blog/2007/06/13/when-does-reuters-use-the-word-terrorist-or-terrorism/ [Accessed March 14, 2008].

31 Maguire.

32 Ariel Merari, "Terrorism as a Strategy of Insurgency," in Gérard Chaliand and Arnaud Blin, eds. *The History of Terrorism: From Antiquity to Al Qaeda* (Berkeley: University of California Press, 2007), 12–51, 16.

33 Norm Goldstein, ed., *The Associated Press Stylebook* (New York: Associated Press, 2007).

34 The Web site organizers believe Hamas and Islamic jihad are terrorist organizations.

35 Honestreporting.com, "Editors Consider the T-word," available online at http://www.honestreporting.com/articles/critiques/Editors_Consider__the_-T-word-.asp [Accessed March 14, 2008].

36 Honestreporting.com.

37 Honestreporting.com.

38 This information comes from personal conversations with a former Iraqi journalist who does not want his name published. He worked for U.S. media and reported on the war for almost five years. He says he still fears for his family's safety because of the time he spent working as a journalist in Iraq. When he talks about differing perspectives on what constitutes terrorism he is speaking mainly about interpretation differences between Sunni and Shia Muslims in Iraq.

39 Committee to Protect Journalists, "Iraq: Journalists in Danger," available online at http://www.cpj.org/Briefings/Iraq/Iraq_danger.html [Accessed April 17, 2008].

40 Bassam Sebti, "Heading into Danger," available online at http://cpj.org/Briefings/2006/DA_spring_06/bassam/bassam_DA.html [Accessed April 17, 2008].

41 El-Nawawy and Iskandar.

42 El-Nawawy and Iskandar, 22 and 24.

43 See, for example, "Suicide Bomber Strikes Iraq Funeral," available online at http://english.AlJazeera.net/NR/exeres/DB389BE0-3597-44BC-8EEBCE1CD04D44C3.htm [Accessed April 17, 2008].

44 El-Nawawy and Iskandar, 41.

45 El-Nawawy and Iskandar, 211.

46 El Nawawy and Iskandar, 100.

47 Shoemaker and Reese.

48 Drew.

49 David H. Weaver and G. Cleveland Wilhoit, The American Journalist: A Portrait of U.S. News People and Their Work (2nd. ed.) (Bloomington, Ind.: Indiana University Press, 1991). See also David H. Weaver and G. Cleveland Wilhoit, The American Journalist in the 1990s: U.S. News People at the End of an Era (Mahwah, N.J.: Lawrence Erlbaum Associates, 1996); David H. Weaver, Randal A. Beam, Bonnie J. Brownlee, Paul S. Voakes, and G. Cleveland Wilhoit, The American Journalist in the 21st Century: U.S. News People at the Dawn of a New Millennium (Mahwah, N.J.: Lawrence Erlbaum Associates, 2007); John W.C. Johnstone, Edward J. Slawski, and William W. Bowman, The News People: A Sociological Portrait of American Journalists and Their Work (Urbana, Ill.: University of Illinois Press, 1976).

50 Weaver and Wilhoit, American Journalist, 1991.

51 Weaver and Wilhoit, American Journalist, 1996, 61.

52 Shoemaker and Reese, 96.

53 Gans, 39.

54 Reynolds and Barnett, "This Just In," 698.

55 Ray Surette, Media, Crime and Criminal Justice: Images and Reality (Belmont, Calif.: Wadsworth, 1998).

56 Robert Lichter, Stanley Rothman, and Linda Lichter, The Media Elite (Bethesda, Md.: Adler & Adler, 1986).

57 Reynolds and Barnett, "This Just In."

58 Media Research Center, "The Full Geraldo Rivera," available online at http://www.freerepublic.com/focus/f-news/572438/posts [Accessed April 19, 2008].

59 James W. Tankard, Jr., Laura Hendrickson, J. Silberman, K. Bliss, and Salma Ghanem, "Media Frames: Approaches to Conceptualization and Measurement" (Paper presented at the annual meeting of the Association for Education in Journalism and Mass Communication, Boston, Mass., 1991).

60 Robert M. Entman, "Framing: Toward Clarification of a Fractured Paradigm," Journal of Communication, 43, (1993): 51–58, 52.

61 James K. Hertog and Douglas M. McLeod, "A Multiperspectival Approach to Framing Analysis: A Field Guide," in Stephen D. Reese, Oscar H. Gandy, Jr., and August E. Grant, eds. Framing Public Life: Perspectives on Media and Our Understanding of the Social World (Mahwah, N.J.: Lawrence Erlbaum Associates, 2001), 139–161, 150.

62 Norris, Kern, and Just, 3–4.

63 Norris, Kern, and Just, 15.

64 Todd M. Schaefer, "Framing the US Embassy Bombings and September 11 Attacks in African and U.S. Newspapers," in Pippa Norris, Montague Kern, and Marion Just, eds. Framing Terrorism: The News Media, the Government, and the Public (New York: Routledge, 2003), 93–112, 93–94.

65 Amy E. Jasperson and Mansour O. El-Kikhia, "CNN and Al Jazeera's Media Coverage of America's War in Afghanistan," in Pippa Norris, Montague Kern, and Marion Just eds. Framing Terrorism: The News Media, the Government, and the Public (New York: Routledge, 2003), 113–132, 129.

66 Jasperson and El-Kikhia, 129.

67 Nacos, 104.

68 Matt Wells, "World Service Will Not Call U.S. Attacks Terrorism," available online at http://www.guardian.co.uk/media/2001/nov/15/warinafghanistan2001.afghanistan [Accessed March 15, 2008].

69 Wells.

70 Wells.

71 Wells.

72 Wells.

73 Entman, 52.

74 Maxwell E. McCombs and Donald L. Shaw, "The Agenda-Setting Function of Mass Media," Public Opinion Quarterly, 36 (1972): 176–187.

75 Walter Lippmann, Public opinion (New York: Macmillan, 1922).

76 Maxwell E. McCombs and Amy Reynolds, "News Influence on Our Pictures of the World," in Jennings Bryant and Dolf Zillmann, eds. Media Effects (2nd ed.) (Mahwah, N.J.: Lawrence Erlbaum and Associates, 2002): 1-18.

77 McCombs and Reynolds.

78 Nacos.

79 Montague Kern, Marion Just, and Pippa Norris, "The Lessons of Framing Terrorism," in Pippa Norris, Montague Kern, and Marion Just, eds. Framing Terrorism: The News Media, the Government, and the Public (New York: Routledge, 2003), 281–302, 290.

80 David H. Weaver, Doris A. Graber, Maxwell E. McCombs, and Chaim H. Eyal, Media Agenda-Setting in a Presidential Election: Issues, Images, and Interest (New York: Praeger, 1981).

81 David H. Weaver, "Political Issues and Voter Need for Orientation," in Donald Shaw and Maxwell McCombs, eds. The Emergence of American Political Issues (St. Paul, Minn.: West, 1977), 107–119.

82 Tamar Liebes and Anat First, "Framing the Palestinian-Israeli Conflict," in Pippa Norris, Montague Kern, and Marion Just, eds. Framing Terrorism: The News Media, the Government, and the Public (New York: Routledge, 2003), 59–74, 62.

83 Werner J. Severin and James W. Tankard, Jr., Communication Theories: Origins, Methods and Uses in the Mass Media (3rd. ed.) (New York: Longman, 1992), 249.

84 The most notable critic was Paul Hirsch. See Paul Hirsch, "The 'Scary World' of the Nonviewer and Other Anomalies: A Reanalysis of Gerbner et al.'s Findings on Cultivation Analysis," Communication Research, 7 (1980), 403–456.

85 George Gerbner, Larry Gross, Michael Morgan, and Linda Signorielli, "The 'Mainstreaming' of America: Violence Profile No. 11," Journal of Communication, 40, 2 (1980): 172–199.

86 Severin and Tankard, 250.

87 Severin and Tankard.

88 Marshall McLuhan, Understanding Media: The Extensions of Man (New York: McGraw Hill, 1964).

89 McLuhan.

90 Severin and Tankard, 251.

91 Kern, Just, and Norris, 282.

92 Kern, Just, and Norris, 282–283.

93 Hoffman, 174.

94 Shoemaker and Reese, 253.

95 Norris, Just, and Kern, 15.

96 Brian Jenkins, "International Terrorism: A New Mode of Conflict," in David Carlton and Carlo Schaerf, eds. *International Terrorism and World Security* (London: Croom Helm, 1975), 16, quoted in Hoffman, 174.

CHAPTER 3
The News Media and the Government

[Former CBS News Anchor] Dan Rather was just a cheerleader for power, a cheerleader for the [Bush] administration. Even now, when he goes to redeem himself with the pictures of Abu Ghraib prison, he sits on the story for two weeks. Nobody's called him on this. Because the chairman of the Joint Chiefs of Staff asked him to sit on the story? And then he says, essentially, "We only broke the story because we had to. We didn't break the story because we should, or because it is the right thing to do for the American people or for the world. We broke the story because somebody else was going to break it. It was going to come out on the Internet." This is just utter cowardice on his part.

       —John MacArthur, Harper Magazine president and publisher[1]

The formal study of the press and its relationship to government dates back to the emergence of the academic field of communication. Studies in the 1940s focused on the impact mass communication had on voters' choices during political elections, and researchers concluded that little evidence existed to show that the media changed people's opinions, even though they did find evidence that people were informed by the mass media.[2]

As chapter 2 showed, a variety of mass communication theories suggest that the media do help set the public agenda and can influence how people perceive information depending on how it is framed and how familiar and relevant an issue is. Two ways that American scholars often consider the relationship between the press and the government is in the context of how much influence government wields on the news media's agenda (and vice versa), and in a First Amendment context, specifically the guarantee of press freedom and access to information. The relationship between the press and the government is relevant to the study of how the press reports about terrorism. As chapter 1 highlighted, most agree that on some level terrorism is a strategy aimed at forcing political change. If media are a primary conduit for news and information about politics, and if media theories are correct in asserting that the media can influence public opinion, both terrorists and counterterrorism efforts by the government need the news media to reach a mass audience. As Brigitte Nacos explained in *Mass-Mediated Terrorism*, the media are one important piece of what she calls the "Triangle of Political Communication," in which three items form a triangle with arrows flowing both ways. The three corners of Nacos' triangle are the mass media, the general public/interest groups, and government officials/decision makers. Nacos argues that terrorists win entrance into this triangle through the mass media, which she says "provide the lines of communication between public offices and the general public."[3] Nacos suggests that despite the "advantage" terrorists have in catching the media's attention because of their "violent deeds," political leaders and government officials "are nevertheless in excellent positions to dominate the news because they are part of one of the cornerstones" in the political communication triangle.[4]

The Power of the President

Several studies have shown that the U.S. president receives the lion's share of American media attention and that his strong presence in the news gives him power to shape public opinion.[5] In addition, as a prominent source of news, some scholars theorize that the president plays an important role in agenda building. Agenda building is a concept related to agenda setting that focuses on the process through which issues become prominent on the media agenda. Agenda building suggests that issues become salient through a manner that is mutually interdependent between policymakers and the news media.[6]

Previous research has shown that the news media use at least 20 percent of White House news releases, and that news coverage that includes official sources, such as government officials, rarely deviates from what official sources say.[7] *Washington Post* reporter Walter Pincus has observed that "When it comes to government, we moved into a PR society a long time ago. Now it's the PR that counts, not the policy. They can make any policy seem to be the right thing or the wrong thing, depending on what information they put out. They understand how we in the media work much better than we understand how they in the government work."[8] Both journalists and scholars have suggested that the increasing sophistication of government communications and PR does impact the White House's ability to affect the media's agenda.

In research that explored the first twelve hours of CNN's breaking news coverage of the September 11 attacks, Reynolds and Barnett found that the network relied almost exclusively on current or former government sources to make sense of the terrorist attack and to frame effectively the initial reporting. The study concluded that in the first few hours of its reporting on the September 11 attacks, CNN created the impression that war was inevitable and necessary to combat the attacks, and that any military response by the United States was justified and supported by the public. This was consistent with the president's position immediately following September 11.[9]

Political scientists and public opinion researchers have explored the factors that lead to support for a president and his policies in the short term following a crisis. One study suggests that the general lack of criticism of the president in the early phase of a crisis leads to heightened public support for a president immediately following a terrorist attack.[10] Others have argued that it's the news media's overuse of administrative sources coupled with a lack of critical media coverage, consistent with Reynolds and Barnett's findings, that can cause a "rally around the flag" effect.[11]

In a recent agenda building and framing study that explored the impact of presidential communications on media coverage of terrorism between 2001 and 2004, Spiro Kiousis and his colleagues found that "in regards to the terrorism meta-issue, the Bush administration, via presidential communication efforts, and the media exhibited a reciprocal relationship not only in setting the agenda of terrorism-related issues, but also in the overall framing of the terrorism issue as well."[12] The findings of the Kiousis study affirm research conducted two decades ago that explored the influence of the president's State of the Union address on the news media's agenda. Comparing the State of the Union addresses of President Richard Nixon (1970), President Jimmy Carter (1978), and President Ronald Reagan (1982 and 1985) to news agendas at the same time, researchers found that only two of the four speeches produced an agenda building effect showing that the press had influenced the president's agenda. The authors of the study concluded that "the results ... demonstrate just how difficult it is to assess the relationship between the president and the press."[13]

Communication scholar Doris Graber has suggested that the relationship between the president and the press goes through cycles.[14] These cycles can result in the president having a great effect on the media agenda, followed by periods where the media are significantly impacting the president's agenda. A new dimension to consider in the contemporary relationship between the press and the president is the advent of twenty-four-hour news. One phenomenon that grew out of the twenty-four-hour news cycle is called the "CNN effect." While no studies have empirically tested the validity of the CNN effect, discussion of the CNN effect by journalists, media scholars, and political scientists arises in connection with the network's coverage of war and terrorism. Dating back to the 1991 Gulf War and initially only associated with CNN, the CNN effect is defined as the effect that live and continuous television coverage of foreign affairs has on war and diplomacy. As Stephen Hess writes,

CNN covered the [first] Gulf War, beaming its signal into foreign and defense ministries all over the world. Dick Cheney, then secretary of defense in the first Bush administration, often acknowledged that he got more timely, relevant information from CNN than he did from U.S. diplomats. Soon ... old and new networks, spawned in the age of new technology, followed CNN's example of providing continuous coverage of dramatic events, at home and abroad. Though it was still called the CNN effect, it included more than just CNN coverage—it meant that the world was now wired, open to instantaneous coverage, and that the coverage affected everyone and everything, including world leaders and their tactics and strategy.[15]

Journalists Danny Schechter and Aliza Dichter write that "there is no denying CNN's enormous impact on world events and the global news agenda. Many governments and news organizations have its electronic face on twenty-four hours a day."[16] In the context of terrorism coverage and war, CNN network executives have acknowledged that people "seek to use us as a player," in much the same way the Irish News of the 1990s was aware of the ways groups like the IRA wanted to "use" them to spread IRA-related paramilitary propaganda.

In a recent study, political scientist Robin Brown explored the "war as a continuation of politics" thinking in the context of the relationship between the military and the media. He notes that "while initial discussion was concerned with the impact of CNN as a diplomatic tool, by the mid-1990s this discussion evolved to ask whether television was actually shaping U.S. foreign policy ... It was argued that the impact of television pictures 'forced' policy makers into taking action whether it was to intervene in humanitarian crises or to pull out once intervention had actually occurred. As with the earlier conflict, an academic literature developed disputing the initial claims. This literature has demonstrated that it is difficult to find evidence for claims about the power of the media over policy."[17]

In his recent book about media framing and public policy, Robert Entman suggests that when political communication scholars examine the "government-media nexus in foreign policy" they use one of two major approaches, hegemony or indexing. Hegemony theorists believe that "government officials keep the information available to the public within such narrow ideological boundaries that democratic deliberation and influence are all but impossible." [18] In other words, the government doesn't allow wide parameters for debate from a wide range of perspectives: rather, it keeps the public's choices very simple and constrained. Indexing suggests that the media "index" or closely reflect debate about policy issues among elites, and do not foster or report about debate as it occurs among regular citizens. Both hegemony and indexing suggest that the media's coverage of foreign policy reflects mostly the government's ideas in very narrow terms, and that this kind of coverage severely limits the public's information and access to a wide range of points of view. The impact of this is that the press "does not offer critical analysis of White House policy decisions unless actors inside the government (most often in Congress) have done so first. This means the media act ... as a vehicle for government officials to criticize each other. Thus the media make no independent contribution (except at the margins) to foreign policy debate."[19]

From a hegemony perspective, while elites may sometimes conflict, their general agreement on basic principles creates a "harmony that impedes the flow of independent information and consistently (although not inevitably) produces pro-government propaganda—and public consent or acquiescence to White House decisions."[20] The concept of hegemony is explored in greater detail in chapter 7. The significance of propaganda to the media coverage of the U.S. "war on terrorism" is discussed below.

Propaganda

The term propaganda originates with the Catholic Church during the Reformation. The Church established the *Congregatio de propaganda fide* (Congregation for the Propagation of Faith) in 1622 to manage its struggle against the growing use of science to better understand the world. A principle figure caught in the struggle was Galileo, who argued that the Earth revolved around the Sun, based on his scientific observations through a telescope. The Inquisition tried and convicted Galileo of heresy in 1633, and he was forced to renounce his view.[21] Mass communication scholars Werner Severin and James Tankard suggest that the term propaganda may have picked up some of its negative connotations of being untruthful from the Galileo incident in which the Church was forced to argue a position that was scientifically false.[22]

Current definitions of propaganda in the field of mass communication trace back to Harold Lasswell's classic book *Propaganda Techniques in the World War*, published in 1927.[23] Lasswell defined propaganda as "the control of opinion by significant symbols, or, to speak more concretely and less accurately, by stories, rumors, reports, pictures and other forms of social communication."[24] Ten years later, Lasswell altered the definition slightly and suggested, "propaganda in the broadest sense is the technique of influencing human action by the manipulation of representations. These representations may take spoken, written, pictorial or musical form."[25] Lasswell's definitions were broad enough to include advertising and other forms of communication that some would call persuasion, so subsequent efforts to refine the term's definition tried to distinguish propaganda from other forms of communication. A modern dictionary defines propaganda as "the spreading of ideas, information, or rumor for the purpose of helping or injuring an institution, a cause or a person" and "ideas, facts or allegations spread deliberately to further one's cause or to damage an opposing cause; also: a public action having such an effect."[26] Still, even today, many would argue that determining whether something is propaganda or not lies in the eyes of a message's receiver as well as in the intentions of the sender.

Most discussion and analysis of propaganda as it relates to the media and government comes from wartime. In 1927, Lasswell noted four main objectives of wartime propaganda—to mobilize hatred against the enemy, to demoralize the enemy, to preserve ally friendships, and to gain the cooperation of neutral parties.[27] Some scholars have noted that the concept of wartime propaganda dates as far back as *The Art of War*, written by Sun Tsu before the beginning of the Common Era. In the United States, most discussion about propaganda begins in connection with World Wars I and II. The U.S. government's highly effective use of propaganda during World War I led many to argue for propaganda education at the onset of World War II. During the years leading up to the U.S. involvement in World War II, Americans worried about the success of Adolf Hitler and his propaganda minister Joseph Goebbels in Germany.[28] In 1939, the newly formed Institute for Propaganda Analysis published the now famous book *The Fine Art of Propaganda*, which explained to readers the seven common devices of propaganda. These devices could be used to identify propaganda and to educate public school students about the dangers of propaganda.[29] The seven devices came from the speeches of Father Charles E. Coughlin, a popular Catholic priest who broadcast his messages over a forty-seven radio station network in the late 1930s. The Institute feared Coghlin could become an "American Hitler" because his shows presented a "fascist" philosophy and his magazine mirrored Nazi propaganda.[30]

The seven propaganda devices are name-calling, glittering generality, transfer, testimonial, plain folks, card stacking, and band wagon.[31] For the past several decades, many people have considered the use of the term "terrorist" as a form of propagandistic name-calling. Some of this traces back to the saying "one man's terrorist is another man's freedom fighter." As Alfred McClung Lee and Elizabeth Briant Lee define it in *The Fine Art of Propaganda*, name-calling is "giving an idea a bad label" and is used to encourage people to "reject and condemn the idea without examining the evidence."[32] As Severin and Tankard point out, "whether a person is called an 'underground' soldier or 'freedom fighter,' or is called a 'terrorist,' depends on the viewpoint of the person assigning the label, or the side the person assigning the label supports. Often the *activities* of an 'underground' soldier or 'freedom fighter' and those of a 'terrorist' are *identical*—only the *label* changes."[33] Dr. Ariel Merari, head of the Center for Political Violence at Tel Aviv University, has suggested that much of the difficulty in defining terrorism grows out of this kind of usage of the term as a general descriptor or label of all negative violent behavior. Merari says that when this kind of generalization occurs, the word terrorist is reduced to a form of propaganda.[34]

Despite attempts to create a distinct and separate category of communication called propaganda, in most of the seven devices, people see linkages to advertising and to political communication. Lee and Lee define "glittering generality" as a way to associate a concept or an idea with a "virtue word," something that will encourage people to accept and approve of the concept or idea without examining the evidence.[35] "Transfer" works through a process of association—the goal is to connect an idea or a cause with something that people like or find favorable. "Testimonial" involves having a highly respected person—or a much despised person—discuss, praise, or denounce an idea to impact people's perceptions.[36] "Plain folks" works when speakers convince audiences that they are one of them, just one of the crowd. "Card stacking" involves the selection of arguments and evidence that support a position while ignoring contrary arguments and evidence. And, "band wagon" works by convincing people to follow the crowd, to embrace an "everybody thinks or does this" mentality.[37]

In addition to name-calling, card stacking and band wagon are the two propaganda devices most used by the government to combat terrorism. A *New York Times* investigation published in April 2008 showed that the George W. Bush administration's communications arm at the Pentagon carefully orchestrated commentary on the Iraq War, terrorism, and treatment of prisoners at Guantanamo Bay by a group of retired military officers who were working for several media outlets as analysts. The *Times* article noted that

To the public, these men are members of a familiar fraternity, presented tens of thousands of times on television and radio as "military analysts" whose long service had equipped them to give authoritative and unfettered judgments about the most pressing issues of the post-Sept. 11 world. Hidden behind that appearance of objectivity, though, is a Pentagon information apparatus that has used those analysts in a campaign to generate favorable news coverage of the administration's wartime performance ... The effort, which began with the buildup to the Iraq war ... has sought to exploit ideological and military allegiances ... [38]

The *Times* article documents an elaborate example of card stacking. Some of the military analysts interviewed for the story acknowledged that their commentary amounted to propaganda. Kenneth Allard, a former military analyst for NBC News, told the *Times* that the government's campaign "amounted to a sophisticated information operation. "This was a coherent, active policy," he said.[39] The *Times* also reported that "A few [analysts] expressed regret for participating in what they regarded as an effort to dupe the American public with propaganda dressed as independent military analysis."[40] The article, which was based on the *Times'* review of records and interviews, noted that the records revealed "a symbiotic relationship where the usual dividing lines between government and journalism have been obliterated."[41]

The effectiveness of propaganda continues to be an area of interest to mass communication scholars. To date, only three of the seven devices—testimonial, band wagon, and card stacking—have been tested experimentally to show that they can be effective on some people.[42] But, the study of all seven devices has become associated with attempts to build a theory of attitude change. One of the first general theories about the mass media—the bullet theory—fits this area of research. The bullet theory suggests that people are "very vulnerable" to mass communication messages and that if a message "hits the target" it will have the desired effect.[43] But, as subsequent theories have shown, the bullet theory is an oversimplification. Severin and Tankard note that while the bullet theory has been revised over time and the seven devices may be limited in their impact, they can still "serve their initial purpose of providing a checklist of techniques commonly used in mass communication. In one way or another, all the propaganda devices represent faulty arguments. Knowledge of the devices can make people ... better consumers of information."[44]

Access to Information

The New York Times article that highlights the power and contemporary use of card stacking also suggests that leveraging access to government information was central to gaining the cooperation of some of the retired military experts. The Times article said that "the Bush administration has used its control over access and information in an effort to transform the analysts into a kind of media Trojan horse—an instrument intended to shape terrorism coverage from inside the major TV and radio networks" and that military commentators who did not support President Bush's positions would suffer for it. "The administration has demonstrated that there is a price for sustained criticism," many analysts said. "'You'll lose all access,' Dr. [Jeffrey] McCausland [CBS military analyst] said."[45] The irony to the government's strong-armed approach is that a significant amount of research suggests that "government officials tend to dominate reporting on foreign and security policy, especially when it concerns military conflict or the likelihood of military deployment."[46] Still, one of the ways that governments control the messages communicated during war and times of crisis, such as during terrorist attacks, is to control the dissemination of information by restricting press access. Although not a First Amendment guarantee, the right of access to information can be found explicitly in statements of the country's founders as well as in state and federal statutes and implicitly in the decisions of the Supreme Court.

An informed citizenry is crucial to a functioning democratic government, and access to information about the workings of the government is key to that process. James Madison noted this when he wrote "Knowledge will forever govern ignorance: And people who mean to be their own Governors must arm themselves with the power which knowledge gives."[47] According to the American Society of Newspaper Editors' Statement of Principles, "The American press was made free not just to inform or just to serve as a forum for debate but also to bring an independent scrutiny to bear on the forces of power in the society, including the conduct of official power at all levels of government."[48] Prominent First Amendment scholars also note the importance of access to information as a foundation for democracy. Alexander Meiklejohn has argued that citizens make "ill-considered" decisions without information; Thomas Emerson has said that to have democracy without an informed public is a contradiction.[49]

Although historical precedent exists, the Supreme Court has not explicitly recognized a First Amendment right to obtain government information. However, judicial statements have hinted at this right. Supreme Court Justice Potter Stewart said that the Constitution is "neither a Freedom of Information Act nor an Official Secrets Act."[50] Justice Byron White wrote in *Branzburg v. Hayes* (1972) that "without some protection for seeking out the news, freedom of the press could be eviscerated."[51] But, the Court has implicitly recognized this right of access when dealing with the public's right to attend trials or receive other information.[52]

The right of access to information is firmly grounded in federal and state statutory law in the United States. Federal access laws developed after World War II, beginning with the Administrative Procedures Act of 1946, which was amended in 1966 to include the Freedom of Information Act (FOIA).[53] The FOIA mandates that all government information generated by executive branch agencies must be disclosed except for material fitting within nine specific exemptions, for example, matters of national security, law enforcement, or personal privacy. However, the exemptions are still framed in favor of disclosure. Nondisclosure is permitted, but it is not mandatory—it remains at the discretion of the agency based on its assessment of risk.[54] Because access rights are statutory, legislatures are often amending the laws, sometimes to benefit certain interest groups or to protect privacy or national security interests.

At the end of 2007, President Bush signed the OPEN Government Act of 2007, which amended several procedural provisions of FOIA, but did not broaden the kinds of information subject to disclosure. It did, however, add some new requirements to what government agencies must do when receiving a FOIA request, including establishing a system that would allow the government to track responses to FOIA requests; responding to requests within twenty days; reporting annually on government agency response times to FOIA requests; reporting annually on the statutory reasons a FOIA request was denied; and requiring better reporting of legal actions against federal agencies that don't comply with FOIA. The amendments also expand the definition of "representative of the media" to include freelance journalists.[55]

What was noticeably absent from the 2007 amendments was any effort to reaffirm the idea that records covered by FOIA are presumptively open unless disclosure would result in harm. This kind of presumption of openness prevailed under FOIA until October 2001, when then attorney general John Ashcroft issued a memo that instructed federal agencies to err on the side of closure if the release of the information requested might harm national security. Ashcroft's FOIA memo said that the Department of Justice (DOJ) is "committed to full compliance with the Freedom of Information Act" but that the DOJ and the Bush administration is equally committed to "protecting other fundamental values that are held by our society."[56]

In addition to protecting national security, Ashcroft listed protection of personal privacy as the other fundamental value that the government sought to protect when it refused in 2001 to release the names of six hundred suspects detained by the United States in connection with the September 11 attacks. Many of the people detained were held on immigration charges; for that reason, Ashcroft said, "it would be inappropriate for us to either advertise the fact of their detention or to provide the suggestion that they are terrorists in a way which would be prejudicial not only to their privacy interest but personal interest."[57]

Some members of Congress, the American Civil Liberties Union (ACLU), and others challenged the Department of Justice's refusal to release the names of detainees. On October 29, 2001, the ACLU and more than a dozen other organizations filed a FOIA request to obtain information about the detainees after repeatedly asking administration officials to disclose the information.[58] When their FOIA request was denied, the ACLU and eighteen other organizations filed suit in federal district court. The complaint asked for the immediate disclosure of government documents concerning the more than a thousand individuals who had been detained or arrested since September 11. Of those individuals, six hundred remained in custody at the time of the filing.[59] In 2002, the district court ordered the release of the names of the detainees and their attorneys under FOIA but said the government could continue to withhold the dates of arrest and release, locations of arrest and detention, and the reasons for the detention under FOIA's Exemption related to law enforcement investigations. In 2003, the U.S. Court of Appeals for the District of Columbia overturned by a 2-1 vote the district court's ruling that the names of the detainees and their attorneys should be released under FOIA. The court also refused to "expand the First Amendment right of public access to require disclosure of information compiled during the government's investigation of terrorist acts."[60] In 2004, the Supreme Court declined to hear an appeal in the case.

Despite the legal outcome of the case, the ACLU and other organizations were not alone in their concern about Ashcroft's decision not to release the names and other information about the detainees. Then assistant attorney general Michael Chertoff appeared before the Senate Judiciary Committee in November 2001 to explain the Department of Justice's response to September 11. Many of the questions he fielded involved the administration's lack of response to requests for information about the detainees and other terrorism-related issues. At that hearing, several senators agreed with the administration's approach to withhold information to protect personal privacy. Utah Senator Orrin Hatch said, "I agree ... that (releasing a list of names) would not only alert our enemies to the status of our investigation, it would also violate the privacy of those being held."[61]

Wisconsin Senator Russell Feingold, who had previously requested detainee information and was denied, told Chertoff:

> I have not found the justifications for not providing the information terribly convincing. I continue to be deeply troubled by your refusal to provide a full accounting of everyone who has been detained and why ... I simply disagree with the attorney general's assertion that disclosing the identities of detainees will bring them into disrepute. I think that just the opposite is true.[62]

In addition to restricting FOIA after the terrorist attacks of September 11, in June 2002 President Bush sent Congress his Homeland Security Act of 2002, which asked for the creation of the Department of Homeland Security. The Act, which was subsequently passed by Congress, excluded the agency from coverage under the Freedom of Information Act. As part of the Department of Homeland Security (DHS), the Protected Critical Infrastructure Information (PCII) Program was created. The PCII is a program designed to restrict access to information provided by the private sector to the DHS related to the country's infrastructure systems (water, energy, computer networking, etc.). According to the DHS, federal, state, and local agencies use PCII to "analyze and secure critical infrastructure and protected systems, identify vulnerabilities and develop risk assessments, and enhance recovery preparedness measures."[63] If the information given to the DHS meets the requirements of the Critical Infrastructure Information Act of 2002, then it is shielded from FOIA, state, and local disclosure laws and from use in civil litigation. Violations of the information protections can result in criminal penalties.

Advocates for access to information have vigorously objected to the secrecy that cloaks the Department of Homeland Security, and journalists and others have suggested that the application of the critical infrastructure exemption has been applied in inappropriate ways. For example, in 2006, the federal government filed a criminal complaint against author, journalist, and film-

maker Greg Palast and television producer Matt Pascarella. The two were filming a story about the plight of Hurricane Katrina evacuees, and their video
showed thousands of New Orleans residents who were evacuated to a make-
shift trailer park about one hundred miles from the city. The temporary hous-
ing was next to a large Exxon Oil refinery, and two weeks after Palast and
Pascarella videotaped the refinery as part of their Katrina story, the government said they had committed a crime—the unauthorized filming of "a critical national security structure."[64]

Conclusion

Brigette Nacos writes that during times of national crisis, "docile media organizations allow presidents and other governmental leaders far more latitude
to carry out their policies than they would in times of normalcy."[65] According
to one researcher, the September 11 terrorist attacks represented a critical
turning point in the relationship between the press and government policymakers that led to the media fully embracing the "war on terrorism" frame
that emerged soon after September 11.[66] Nacos agrees:

> In the nearly four weeks since the terror attacks of September 11th, the war metaphor
> had been invoked so often by media organizations and public officials that the Ameri-
> can public was hardly surprised when President Bush revealed the start of the military
> phase in what cable TV networks had long described in their on-screen banners as
> "America's New War" or "War against Terrorism."[67]

Since September 11, the public saw increases in terrorism-related propaganda issued by the government and restrictions on the public's access to information, neither of which the news media significantly challenged. The one
exception to this came with the embedding of the media within U.S. troops in
Iraq. Since the Vietnam War, the U.S. government has resisted giving little if
any media access to U.S. military operations in the field. Government leaders
have blamed the news media for turning the public against the war effort in
Vietnam and have struggled to keep the media away from armed conflict for
decades.[68] Nacos calls the government's reversal of position and the resulting
"large-scale embedding" of the news media during the Iraq invasion a "stroke
of genius and a complete success from the perspective of the Bush administration and the U.S. military."[69]

One can get a glimpse of the military's perspective on embedding by
reading its guidelines on embedded media. According to *The U.S. Army & Marine*

Corps Counterinsurgency Field Manual, "the media are a permanent part of the information environment. Effective media/public affairs operations are critical to successful military operations."[70] The manual says that "Embedded media representatives experience Soldiers' and Marines' perspectives of operations in the [counterinsurgency] environment. Media representatives should be embedded for as long as practicable." It adds that

> the media are ever present and influence perceptions of the [counterinsurgency] environment. Therefore, successful leaders engage the media, create positive relationships, and help the media tell the story. Operations security must always be maintained; however, security should not be used as an excuse to create a media blackout. In the absence of official information, some media representatives develop stories on their own that may be inaccurate and may not include the [counterinsurgency] force perspective.[71]

The success of embedding journalists within the troops fighting in Iraq was designed to expand media access, but was also intended to deliberately further the U.S. cause in the "war on terrorism." Is this propaganda, access, or both? About five hundred journalists were embedded within U.S. and British military units during Operation Iraqi Freedom in 2003, and another two thousand "unilateral" journalists were in Kuwait. Victoria Clarke, the Assistant Secretary of Defense for Public Affairs at the time, said that, "it's in our interest to let people see for themselves through the news media, the lies and deceptive tactics Saddam Hussein will use."[72] The government approved its embedded media policy after journalists complained about the lack of access to the war in Afghanistan. In recalling her experiences reporting from Afghanistan in 2001, *Washington Post* reporter Carol Morello tells the story of a friendly fire incident that resulted in both American and Afghan casualties. Morello said:

> So the reporters were like hungry dogs. What can we do? A photographer says, "Can I go take pictures?" He's told no. One of the print reporters says, "Can we at least go stand there and watch?" We're told no. The public affairs officer says, "What, you want to go see dead Americans?" We said, "No, we think this is our job." We said, "Can we talk to the pilots who flew them back?" No. "Can we talk to any of the medics when they're done?" "Well, they're too tired." "Can we talk to any of the Afghans who have minor injuries?" "No, we don't have a translator." News was breaking 100 feet from where we were standing and we were ordered to stay in that warehouse. We did not get any of it.[73]

Prior to the emergence of the formal embedding policy in 2003, some journalists did get unrestricted access to the troops, and Clarke suggested that the government saw the value in this: "A fellow from a news organization I will not name ... spent two or three weeks with the marines near Kandahar. Two or three weeks sleeping on the ground with the bugs, the cold, everything else, and he said it was extraordinary. And to the extent [that] you can facilitate more of this, do it."[74]

Whether it's carefully orchestrated propaganda or simply providing access to fighting a "war against terrorism," the government's strategy does come with some risks. As Vietnam showed, the importance of public opinion to the maintenance of support for any war can sometimes hinge on the media portrayals of the war, particularly the images. As Nacos notes, "at the heart of this enduring Vietnam War legacy among military and civilian leaders is the belief that the mass media—especially television—will turn domestic public opinion against involvement in military conflicts by dwelling on visual images of the ugly side of war. The related 'body bag thesis' holds that the contemporary public cannot stomach casualties."[75] Chapter 4 will explore the power of the image in the coverage of recent terrorism events, and chapter 5 will highlight the special role that television plays.

Notes

1   John MacArthur, "Everybody Wants to Be at Versailles," in Kristina Borjesson, ed. *Feet to the Fire: The Media after 9/11* (Amherst, N.Y.: Prometheus Books, 2005): 92–122, 105.

2   Paul Lazarsfeld, Bernard Berelson, and Hazel Gaudet, *The People's Choice* (New York: Columbia University Press, 1944).

3   Brigette Nacos, *Mass-Mediated Terrorism: The Central Role of the Media in Terrorism and Counterterrorism* (Lanham, Md.: Rowman & Littlefield, 2007), 15.

4   Nacos, 197.

5   Doris Graber, "Introduction: Perspectives on Presidential Linkage," in Doris Graber, ed. *The President and the Public* (Philadelphia: Institute for the Study of Human Issues, 1982): 1–14; Roy L. Behr and Shanto Iyengar, "Television News, Real-World Cues, and Changes in the Public Agenda," *Public Opinion Quarterly*, 49 (1985): 38–57.

6   Spiro Kiousis, Michael Mitrook, Trenton Seltzer, Cristina Popescu, and Arlana Shields, "First- and Second-Level Agenda Building and Agenda Setting: Terrorism, the President and the Media" (Paper presented to the International Communication Association, June 16, 2006, Dresden, Germany).

7   Judy VanSlyke Turk, "Between President and Press: White House Public Information and Its Influence on the News Media" (Paper presented to the Association for Education in Journalism and Mass Communication, August, 1987, San Antonio, Texas); Leon V. Sigal, "Sources Make the News," in Robert Manoff and Michael Schudson, eds. *Reading the News* (New York:

Pantheon, 1986), 9–37.

8   Walter Pincus, "Guerilla at the Washington Post," in Kristina Borjesson, ed. *Feet to the Fire: The Media after 9/11* (Amherst, N.Y.: Prometheus Books, 2005), 218–248, 221.

9   Amy Reynolds and Brooke Barnett, "'America under Attack': CNN's Verbal and Visual Framing of September 11," in Steven Chermak, Frankie Y. Bailey, and Michelle Brown, eds. *Media Representations of September 11* (Westport, Conn.: Praeger, 2003), 85–101.

10 George Shambaugh and William Josiger, "Fear Factor: The Impact of Terrorism on Public Opinion" (Paper presented to the International Studies Association [ISA] Annual Convention, Chicago, Ill. February, 2007), 9.

11 John Mueller introduced the phrase "rally-round-the-flag" to describe the sudden and relatively short-lived increases in a president's approval ratings following unexpected, high-profile international events. John E. Mueller, *War, Presidents and Public Opinion* (New York: John Wiley and Sons, 1973). See also Shambaugh and Josiger. Chapter 7 explores the impact of patriotism on press coverage of terrorism in detail.

12 Kiousis et al., 19.

13 Wayne Wanta, Mary Ann Stephenson, Judy VanSlyke Turk, and Maxwell E. McCombs, "How President's State of Union Talk Influenced News Media Agendas," *Journalism Quarterly* (1989): 537–541, 541.

14 Doris Graber, *Mass Media and American Politics* (Washington, D.C.: Congressional Quarterly, 1984).

15 Stephen Hess and Marvin Kalb, eds., *The Media and the War on Terrorism* (Washington, D.C.: Brookings Institution Press, 2003), 63.

16 Danny Schechter, *Media Wars: News at a Time of Terror* (Lanham, Md.: Rowman & Littlefield, 2003), 46.

17 Robin Brown, "Clausewitz in the Age of CNN: Rethinking the Military-Media Relationship," in Pippa Norris, Montague Kern, and Marion Just, eds. *Framing Terrorism: The News Media, the Government and the Public* (New York: Routledge, 2003), 44.

18 Robert M. Entman, *Projections of Power: Framing News, Public Opinion and U.S. Foreign Policy* (Chicago: University of Chicago Press, 2004), 4.

19 Entman, 4. Entman quotes Jonathan Mermin, *Debating War and Peace: Media Coverage of U.S. Intervention in the Post–Vietnam Era* (Princeton: Princeton University Press, 1999).

20 *Ibid.*

21 Werner J. Severin and James W. Tankard, Jr., *Communication Theories: Origins, Methods and Uses in the Mass Media* (3rd ed.) (New York: Longman, 1988).

22 Severin and Tankard, 90.

23 Harold Lasswell, *Propaganda Technique in the World War* (New York: Peter Smith, 1927).

24 Lasswell, 9.

25 Lasswell quoted in Severin and Tankard, 91.

26 *Merriam-Webster's Online Dictionary*, available at http://www.merriam-webster.com/dictionary/propaganda [Accessed April 30, 2008].

27 Lasswell, 195.

28 Severin and Tankard, 92.

29 Alfred McClung Lee and Elizabeth Briant Lee, eds., *The Fine Art of Propaganda: A Study of Father Coughlin's Speeches* (New York: Harcourt, Brace, 1939).

30 Severin and Tankard, 92.

31 Lee and Lee.

32 Lee and Lee, 26.

33 Severin and Tankard, 95. Emphasis in original.

34 Ariel Merari, "Terrorism as a Strategy of Insurgency," in Gérard Chaliand and Arnaud Blin, eds. *The History of Terrorism: From Antiquity to Al Qaeda* (Berkeley: University of California Press), 12–51, 13.

35 Lee and Lee, 49.

36 Lee and Lee, 67, 74.

37 Lee and Lee, 95, 105.

38 David Barstow, "Message Machine: Behind TV Analysts, Pentagon's Hidden Hand," available online at http://www.nytimes.com/2008/04/20/washington/20generals.html?_r=2&oref=slogi&oref=slogin [Accessed May 1, 2008].

39 Barstow.

40 Barstow.

41 Barstow.

42 Severin and Tankard, 105.

43 Severin and Tankard, 105–106.

44 Severin and Tankard, 106.

45 Barstow.

46 Nacos, 143.

47 Gaillard Hunt, ed., *The Writings of James Madison* (New York: G.P. Putnam's Sons, 1910), 103.

48 American Society of Newspaper Editors, "ASNE Statement of Principles," available online at http://www.asne.org/index.cfm?ID=888 [Accessed May 1, 2008].

49 Alexander Meiklejohn, *Free Speech and its Relation to Self-Government* (New York: Kennikat Press, 1948); Thomas Emerson, *The System of Freedom of Expression* (New York: Vintage Books, 1970).

50 Stewart Potter, "Or of the Press," *Hastings Law Review*, 26 (1976): 631–636, 636.

51 *Branzburg v. Hayes*, 408 U.S. 665 (1972), 682.

52 *Richmond Newspapers v. Virginia*, 448 U.S. 555 (1980); *Red Lion Broadcasting Co. v. FCC*, 395 U.S. 367 (1969).

53 *Freedom of Information Act*, 5 U.S.C, sect. 552 (1988).

54 5 U.S.C. 552 (b)(6-7) (C) (1988).

55 Douglas Lee, "What's on the Horizon," available online at http://www. firstamendment-center.org/press/information/horizon.aspx?topic=FOI_horizon [Accessed May 1, 2008].

56 John Ashcroft, *Memorandum for Heads of all Federal Departments and Agencies*, available online at www.usdoj.gov/oip/foiapost/2001foiapost19.htm [Accessed May 1, 2008].

57 John Aschroft, *U.S. Department of Justice News Conference with Attorney General Ashcroft*, available online at usinfo.state.gov/topical/pol/terror/01112711.htm [Accessed May 1, 2008].

58 Nadine Strossen, *Forum on National Security and the Constitution*, available online at www.aclu.org/congress/1012402a.html [Accessed May 1, 2008].

59 *Center for National Security Studies et al. v. Department of Justice* (D.C. Dist. Court, December 5, 2001) (Complaint available online at www.aclu.org/court/detainee-foia-complaint.PDF).

60 *Center for National Security Studies et al. v. U.S. Department of Justice*, 356 U.S. App. D.C. 333 (2003). For a collection of legal documents related to this case, see http://www.cnss.org/cnssvdoj.htm.

61 Capitol Hill Hearing, "Preserving Freedoms while Defending against Terrorism: Panel 1 of a Hearing by the Senate Judiciary Committee," *Federal News Service* (November 28, 2001): 1–49, 6.

62 Hearing, 23-24.

63 Protected Critical Infrastructure Information (PCII) Program, available online at http://www.dhs.gov/xinfoshare/programs/editorial_0404.shtm [Accessed May 1, 2008].

64 Greg Palast, "Palast Charged with Journalism in the First Degree," available online at http://www.gregpalast.com/palast-charged-with-journalism-in-the-first-degree/[Accessed May 1, 2008].

65 Nacos, 143–144.

66 Na'ama Nagar, "Frames That Don't Spill: The News Media and the War on Terrorism" (Paper presented to the International Studies Association [ISA] Annual Convention, Chicago, Ill., February 24, 2007).

67 Nacos, 159.

68 Nacos, 173; see also Daniel Hallin, *"The Uncensored War": The Media and Vietnam* (New York: Oxford University Press, 1986).

69 Nacos, 173.

70 *The U.S. Army & Marine Corps Counterinsurgency Field Manual* (U.S. Army Field Manual No. 3-24, Marine Corps Warfighting Publication No. 3-33.5) (Chicago: University of Chicago Press, 2007), 164.

71 *Counterinsurgency Manual*, 165.

72 Tammy Miracle, "The Army and Embedded Media," *Military Review* (September–October 2003), available online at http://findarticles.com/p/articles/mi_m0PBZ/is_5_83/ai_111573648 [Accessed May 1, 2008].

73 Hess and Kalb, 167–168.

74 Hess and Kalb, 109.

75 Nacos, 164.

CHAPTER 4

The Image

Photographs have the kind of authority over imagination to-day, which the printed word had yesterday, and the spoken word before that. They seem utterly real.

—Walter Lippmann (1922)

Perhaps it was the harsh realness of the photo that caused the controversy after *The Oregonian* newspaper ran a front-page photo in April 2004 of an Iraqi man sobbing over his six dead children, his wife, two brothers, and his parents—killed in an American air bombing. This photo was probingly displayed on the front page on the same day that the newspaper reported the rescue of Jessica Lynch and the death of an Oregon soldier.

The Oregonian received an onslaught of criticism; more than fifty people called, wrote, or e-mailed their complaints. Here are some of the e-mail complaints forwarded to the Ombudsman for the newspaper, as reprinted in the journalism online resource Poynter.org:

Why did you run on front page April 2 LARGE picture of Iraqi family bombed? Instead of the rescued woman soldier as feature? Steven Dean's editorials are shameful. I'm ashamed of Oregon's largest daily newspaper; NOT fair and un-balanced during this war.

I wish to express my deep frustration with the very large picture above and below the fold on the front page of the paper today. I feel that the stories of a rescued soldier and an *Oregonian* killed while in service to his country deserve more space than a narrow left hand column next to this emotional[ly] charged picture. Let me say that I fully appreciate the grief and terror of the Iraqi people. In fact my class and I pray for them daily, but this picture was still not appropriate in light of recent events mentioned above.

I found the picture on today's (4/2/03) *Oregonian* to be a great attempt to give sympathy and support to a merciless enemy. Certainly the story of civilian casualties needs to be told, but not as the main focal point of a daily newspaper. How about celebrating the return of one of our own? Or the tremendous effort we are taking in not harming civilians at the risk of our own? Or some indication of the 10s of thousands of innocents that have been killed by Iraq's own leadership?

I am sickened by your front page today! How can you put that poor G.I. that died in a side column, yet make the Iraqi man who lost his family the main theme? Are you not American? You can bet I will be speaking with my husband this evening, and may be canceling our subscription. Perhaps you should reconsider your stance on producing newspapers—it does NOT look as if you are objective, but ANTI-AMERICAN, ANTI-WAR, yet PRO-PROTESTORS!!'"[1]

Photographs and moving images are the most effective ways to convey emotion. They are rife with cognitive and emotional implications; the public makes decisions about news coverage based on the images. In this case, one image overpowered all the words written and suggested to many readers that the paper was biased.

Visual Bias and Graphic Imagery

The vast majority of studies on visual framing have analyzed media coverage during crisis. Susan Moeller argues that still photography provides a key influence in the public's perceptions of war, for example.[2] Several studies have documented the information that visuals contain and have demonstrated that visual images frame crisis in a way that hinders the public's ability to understand the issues at hand. Images can reduce the Gulf War coverage in 1991 to depictions of war technology[3] and September 11 coverage to a pro-American argument for war.[4]

Television's primary vehicle to convey information is the image, which is rich with emotion-laden nonverbal information.[5] Negative images, such as pain and human suffering associated with crisis, have been shown to be more memorable than other images in television news.[6] It was this idea that sparked CNN to remind reporters to show the video clipping of the burning World Trade Center towers as a constant reminder of why the United States was at war in Afghanistan. The potential emotional impact of images is also the reason for the hot debate.

In October 2001, the chairman of CNN sent a memo to his staff asking them to balance images of civilian devastation in Afghan cities with reminders that the Taliban harbors the terrorists responsible for September 11. CNN Chairman Walter Isaacson wrote to his international correspondents and said, "As we get good reports from Taliban-controlled Afghanistan, we must redouble our efforts to make sure we do not seem to be simply reporting from their vantage or perspective. We must talk about how the Taliban are using civilian shields and how the Taliban have harbored the terrorists responsible for killing close to 5,000 innocent people."

This memo came during a time when errant U.S. bombs had landed in residential areas, at one point even damaging a Red Cross warehouse. Television images of the damage resulting from these bombs fueled criticism of the

American war effort in Afghanistan, a war effort that found 90 percent public support initially. NBC News Vice President Bill Wheatley said, "I'd give the American public more credit, frankly. I'm not sure it makes sense to say every single time you see any pictures from Afghanistan, 'This is as a result of September 11th.' No one's made any secret of that."

By far the most contentious aspect of the visual coverage of the September 11 attacks was the choice of whether or not to show people jumping and falling from the towers, and the body parts strewn at Ground Zero. USA Today—whose editors used eyewitness accounts, forensic evidence, and news video—estimated that at least two hundred people died by jumping—that is, about 8 percent of those who died in New York City on September 11, 2001 died by jumping out of the buildings. Considering the magnitude of this aspect of the story, it received very little coverage. One of the most famous photos that captured the people who fell or jumped from the World Trade Center was taken by Richard Drew and ran on page 7 of The New York Times and in newspapers all over the United States. It ran only once. Newspapers across the country were forced to defend charges that they exploited a man's death and stripped him of his dignity in these last moments of life by publishing this photo. On this day saturated with graphic photos and images, the falling bodies became the only taboo image. Early on in the coverage, CNN showed the bodies falling. At one point, the camera poised on top of a building zooms in on something falling from the tower and appears to hold it long enough to realize the falling object was a person and then zooms back out. CNN later showed only the blurred images of falling bodies as to not be recognized by a loved one. Then, CNN stopped showing them all together. In commemorations of the attacks, the images of falling people were either left out or reduced to just one image from a distant focal point. These images are still readily available, but now on sites that featured autopsy photos of Nicole Brown Simpson and videotapes of executions.[7]

In addition to not showing images of people jumping, traditional media also began to show signs of competing with online sources. Today, CNN may choose not to air graphic video of Former Prime Minister Benazir Bhutto's assassination, but CNN viewers can see that footage online. CNN and other online news sources must grapple with the viewers who would be upset if that video aired as well as those who realize that the network is sanitizing the images and might lose trust in the network to say what is going on. Part of giving the public the information they need is showing the truth, the aptly named awful truth. At times this is bloody and graphic and not pro-American. Because television news is subject to notions of taste and decency and to presumptions about the sensitivities of audiences, it is greatly limited in the types of stories covered and the depth to which they are covered. However, online and print sources often have inside page print space or unlimited online space to post such images.

After Paul Johnson Jr.—a U.S. citizen working for Lockheed Martin in Saudi Arabia—was taken hostage and then executed, the Drudge Report posted photos of the body and severed head. The Pittsburgh Tribune-Review published three less graphic photos of Johnson's body along with a statement from the terrorist on an inside page of the newspaper. Editor Frank Craig argued in an editorial three days after publishing the photos that these images "demonstrate the brutality, the inhumanity, and the deadly danger" of the terrorist.[8] Craig said that Americans should not avert their eyes to the gravity of the dangers in Iraq. He ended the editorial with, "The photos published in Saturday's edition should (emphasis his) offend and horrify you. I hope they also help you to comprehend the enemy's mentality and the extent of the danger we face." Craig also addressed the decision to publish two images on page 1, though no photos of the actual beheading, after American Nicholas Berg was murdered. He wrote, "I felt the images we used showed that act's savagery as adequately as a more graphic image would have done."

The Boston Phoenix touched off a media maelstrom when it published on its Web site a link to a four-minute video created by the terrorists who executed Wall Street Reporter Daniel Pearl. The video showed Pearl forced to talk about his Jewish background while images of Palestinian suffering were shown on screen. Then, after a quick fade-out, Pearl's lifeless body is seen on screen as one of the executioners hacks his head off and holds it up for the camera. Pearl's father weighed in on the decision to publish this photo in an opinion piece for the New York Times where he pleaded with American news media "to preserve the dignity of our champions, we should remove all terrorist-produced murder scenes from our Web sites and agree to suppress such scenes in the future." Other journalists condemned the publication of photos or videos of Pearl's death. However, some argued that journalists who chose not to print or air the images afforded a courtesy to a fellow journalist, which would not be given to others. Further, the video was already available on Web sites that specialized in displaying disgusting photos. News organizations put the images in the proper context. Journalist Dan Kennedy offered this explanation:

> Nor was there anything unusually grotesque about the images when seen in the context of other horrifying news photos, some of them Pulitzer Prize–winners. From the Holocaust to the Vietnam War to the body of a dead American soldier being dragged through the streets of Mogadishu, news photographers have shown us death and destruction in the rawest form imaginable. The fact that the Daniel Pearl video was produced by terrorists rather than journalists is a mere detail. After all, it depicts what happened, which is the most elemental definition of news.[9]

But this defense of the photos might ignore the very real possibility that showing graphic images did not aid in the audience learning about this significant issue. On a CNN show that examined the role the press played in showing graphic images and terrorism, author Jason Burke said:

> And to my mind, it's better to have images, even grisly images, presented in an authoritative way, with analysis around it that gives them some context, that explains the situation, explains who's behind this attack, explains perhaps the reaction of people to that attack, which often the print media can do in a way that is more valuable than the TV media because we have that much more space and time. That way, you put the instant in its context. You're not propagandizing, you're not playing into the hand of the terrorist.

CNN host Nic Robertson responded:

> Do you think the audience gets that? I mean, my impression, Roger, is that the audience looks at this and it's horrified by the images, and they perhaps don't get all the detail that we, as journalists, want to deliver to them around this.[10]

Robertson suggests something that communications scholars have studied quite extensively—the ways that images might hinder the information that journalists intend to convey. Although research does show that pictures generally aid the understanding of issues and people pay more attention to the images than the words, images are limited in terms of how sophisticated the message received can be.[11]

Images and Cognition

Research in the area of negative video has shown that images that are radically different from what we would expect require further processing and that novel images compel viewing.[12] Thus, the increase in cognitive capacity required to process these unusual images remains, and this increases recall of subsequent images. So when people are seeing something they have not seen before, they pay attention and remember it. In this way, Jason Burke's response on CNN about putting the image into context makes sense. The audience gets key information about the terrorist attack from the image as well as the caption and story, rather than potentially relying on the misinformation already stored in their heads. "Furthermore, the images people recalled and the emotions they felt at the time of the September 11 attack may be used to evaluate the level of concern individuals have with terrorism. In other words, the memory of these visuals could have an agenda-setting effect on viewers."[13]

Research also shows that emotional experience is cognitively demanding and absorbing. Those who are preoccupied with emotion will not be able to remember the next story, but after they break out of the preoccupation, they'll have superior information processing.[14] The results are mixed on whether emotionally compelling images divert the spectator from the narrative. Barrie Gunter found that highly emotional visual images have the potential to distract from the narrative; others found memory to be better after emotional images, while still others found that memory was better for textual information but not for visual information. It seems that storage and thus memory depends on the motivations of the viewer and the emotion of the video.[15]

Newhagen and Reeves found that memory is worse for information that preceded the negative scenes but better for the scenes that elicited the negative response.[16] Newhagen also found that negative images inhibit memory for preceding information and enhance memory for material that follows them.[17] He found that factual information and topics were remembered better for stories without compelling images, and visual images were remembered for stories with compelling images. So, it seems that news stories would be more effective if the compelling images were shown first followed by the facts. However, other scholars found that emotional recall was poorer for about three minutes after an emotionally charged disturbing news story. Previous research has hinted that after viewers break out of the preoccupation with the emotionally arousing material they'll have superior information processing for some time, but other scholars have found that the processing simply returned to normal.[18]

These studies suggest that for terrorism stories, placement of the story and negative image have an impact on what is remembered. Perhaps most specific to terrorism coverage is Newhagen's 1998 study of images that induce anger, fear, and disgust. Because terrorism news is often made up of these negative images and the research on memory for negative images has been mixed, Newhagen decided to divide negative or emotion-laden images into more specific categories and look at memory and approach avoidance—the idea that when a novel object enters the environment, is it welcomed or avoided? He found that the most memorable and most approachable images are in the following order: anger, fear, and disgust. So it seems that images of disgust may simply be avoided, and in the case of TV news the audience might change the channel. So if producers want to draw attention to a story, they need to draw attention with images that cause anger and then introduce what they want viewers to remember right after those images.

Graphic images might cause the outrage to suggest something needs to be done, but only as long as those images are novel, something useful for those trying to compel action from the audience. Once the novelty is gone, then compassion fatigue sets in. So the goals of getting a pubic behind a war on terror policy might easily be met with emotional imagery. But for the news professional interested in informing the public, the results are more dismal. These images have the potential to invoke an avoidance response and also have a negative impact on memory. Thus, graphic images may cause the viewers to disengage with the topic and not recall the information presented with it. Further, people afraid of what they see on the news, something that can happen with graphic images, are less able to learn from what they watch.[19]

Television in particular faces tough questions about the images used in terrorism coverage because of the tendency to replay compelling footage despite the lack of new story developments. Footage shown on twenty-four-hour cable news is shown over and over until a new news cycle arrives. The extensive and continuing coverage of terrorist incidents can also lead to intense anxiety in viewers.[20] Combine this tendency with the previous research about stories that follow such footage, and the press is setting up a news cycle that might be avoided, will likely not be remembered, and will induce anxiety in the audience.

Research has also shown the potential effect of graphic images in terrorism coverage. Only negative graphic images affect perceptions of risk; reassuring images and stories have no significant effect on citizen's perceptions of risk.[21] Threatening images in terrorism coverage in particular have a powerful effect on the public. For example, when terrorism coverage increases, so does the public's anxiety and the president's approval rating, according to Brigitte Nacos.[22] Some scholars have looked specifically at how the visual imagery in terrorism coverage affects attitudes toward governmental leaders and policies and found that exposure to these images increases the approval ratings of the president, something called the "September 11 halo."[23] Another scholar found that these reactions are mediated by partisanship, indicating that media coverage is received differently depending on your response to the war on terror.[24] Another study showed that when the audience reacted to images of September 11 with sorrow, shock, or worry, they were more likely to recall several images from the coverage. This study showed once again that emotion influences recall.[25]

Despite ethical and recall issues of graphic images, the news is fairly good for the use of images in general. Research has shown that pictures generally aid recall. People remembered the pictorial items more often than the nonpictorial ones.[26] It seems that high imagery conditions are effective in learning and memory because they increase the probability that both the verbal and visual content will be coded. This helps with recall because there are two things to retrieve from memory in case one is forgotten.[27]

When there is a disconnect between the image and the words, the viewer resorts to taking in the information from the image rather than the words, a medium that conveys less sophisticated information.[28] In this way, Burke's response on CNN about putting the graphic image into context makes sense. The audience gets key information from the information surrounding the image. Newspaper and TV images also convey subtle forms of patriotism, perhaps the type of images the readers of the Oregonian were looking for instead of the photo demonstrating the horror to the Iraqi people. These types of photos celebrate the greatness of the United States and the war effort. Editors at newspaper choose among a lot of photos that are at their disposal. Coverage of September 11 can show either the iconic image of September 11 as Iwo Jima or can show the burning towers. One image is chaos and the other invokes symbols of conquering. These images framed the issue of September 11 for the reader, and similar selection of images frame other terrorist events.

Shana Kushner Gadarian argues that visual images during terrorism coverage have an influence not only on emotion but also on policy: "Pictures of the World Trade Center or bloodied victims of a terrorist attack may arouse various emotions into the 'right' policy to fight terrorism. The Bush administration consistently uses allusions and images of September 11 to argue that the

United States must employ an offensive foreign policy that seeks out terrorists abroad in order to protect its citizens at home."[29]

Scholars have debated this so-called CNN effect, the way flash news affects foreign policy decisions; such an effect includes President George H. W. Bush's sending of troops to protect the Kurds in northern Iraq in 1991 when television showed their suffering. Another classic example given is Americans pulling out of Somalia when television showed an American soldier's body being dragged through the streets of Mogadishu. But others dismiss this claim and say the CNN effect is more a function of weak journalism that too closely follows the government's lead.[30]

Journalist Robert Fisk argues about the unfair media treatment in the Is-raeli-Palestinian conflict and cites a cover of Newsweek magazine on February 19, 2003 that showed a Palestinian man in a kuffiah headdress holding an automatic rifle under the headline "Terror Goes Global" in reference to Osama bin Ladin's terror network. Even though the reader was led to believe that this man was a terrorist, Fisk thought he'd seen this photo before and that something was awry. He called up the photographer and confirmed that the man in the photo was at a Palestinian funeral. Fisk explained:

> Thus a Palestinian gunman armed and attending the funeral of a fellow Palestinian killed by Israelis had been turned into a representation of "global terror." Palestinians as people—and the man on the cover was definitely Palestinian—had been effortlessly transformed into enemies of the world. It wasn't the photographer's fault, but News-week's cover picture was a lie. The man whose face was covered by the kuffiah—dangerous though he would be to Iraelis—had nothing to do with bin Laden or the lead story in the magazine.[31]

Bridgette Nacos writes about the impact of images during coverage of the attack of the World Trade Center. She writes, "The mass-mediated visuals of the hard hit and disabled warship, the pictures of the victims and their families, and the images of frustrated Washington officials were the message that revealed more about the motives than any eloquent claim of responsibility."[32] Nacos argues that even when terrorists do not claim responsibility, their message is still communicated as the press covers the attack.

The way that journalists described the situation combined with the images shown also creates a powerful message. Standley, a BBC correspondent in New York City, said the following as images of the burning towers aired:

Americans tend to get to work early, it happened at 9 o'clock local time here in New York, so many of those people would have been at their desks. The whole area around the Financial District, there's lots of subway lines, very high, big office buildings, all of them heavily populated with people. The pavements down there are just packed with people, they would have been injured; there is no way of avoiding that condition. Some people would have been killed surely when debris was falling 110 floors at the world Trade Center, when the planes collided with it.

For example, the images used in the coverage of the 2005 London bus and underground bombings as well as the accompanying description gave the viewer a clear picture of the victim and the terrorist. The earliest images of the attacks on July 7, 2005 were quite low tech, most coming from cell phone cameras of eyewitnesses. The BBC imagery consisted of a stationary wide-angle image that mostly included rescue vehicles and very little movement. The graphic coverage did not come from actual images but rather eyewitness descriptions of the events, employing graphic descriptions; CNN Reporter on July 7, 2005 describing an eyewitness account:

He said that people were praying out loud. It was pitch black. The smoke was filling the air. People were banging against the windows, trying to get out of the Tube, trying to get rid of the smoke. People thought, and he thought indeed that he was going to die. He thought, he said, his time was up. People trying to get out all the time, screaming, hysteria. It was like a scene from hell.

London Times, July 12:

A decapitated head was found at the bus scene which has been, in Israeli experience, the sign of a suicide bomber.

The cover images on newsmagazines after September 11 suggested the media frames that would be employed surrounding the terrorist attacks. Many magazines showed photos of the burning towers. Magazines that did not feature news photos on the cover chose simple graphic representations of the events, such as the edition of the New Yorker released on September 18, 2001, which featured a black cover with September 11, 2001, written in red vertically down the cover. The images chosen ascribed different meaning to the day, whether the focus was on triumph or sorrow. During early coverage of the September 11 attacks, the images focused on the planes crashing into the World Trade Center Towers. But soon the focus shifted to one iconic image that appeared on the September 24 Newsweek cover, that of three firemen with an intact flag pole rising from the rubble of the trade center. This image immediately harkened to Iwo Jima and the Pulitzer Prize–winning photo. Both of these images suggest the ultimate victory of the United States over its enemies, with the headline "God Bless America."

This photo became the symbol of the attacks. It was shown during the president's speech at the National Cathedral memorial service three days after the attack. This photo was also left behind during a U.S. forces raid of a Taliban headquarters in Afghanistan. The image was made into a commemorative coin. This image was the one that the U.S. government preferred to show in summing up the attack, the iconic image of triumph over tragedy.

First Images of September 11, 2001

Looking more deeply into one terrorist event through a qualitative frame analysis illustrates the ways that images convey and frame the dominate message of the coverage.

Scholars have conducted frame analyses on interpersonal and mass communication messages. Sociologist William Gamson has suggested that a complete frame analysis has three components: an examination of the production process that "alerts us to issues of power and resources";[33] an examination of texts, which can occur on different levels of analysis; and an attempt to address the "complex interaction of texts with an audience engaged in negotiating the meaning."[34] Robert Entman wrote "to frame is to select some aspects of a perceived reality and make them more salient in a communicating text, in such a way as to promote a particular problem, definition, causal interpretation, moral evaluation, and/or treatment recommendation for the item described."[35] Entman argued that the social power of journalism is reflected in the ways that journalists designate what is most salient through, among other things, the sources they select and the ways in which they provide order and context to stories:

> Communicators make conscious or unconscious decisions in deciding what to say, guided by frames ... that organize their belief systems. The text contains the frames, which are manifested by the presence or absence of certain keywords, stock phrases, stereotyped images, sources of information, and sentences that provide thematically reinforcing clusters of facts or judgments.[36]

James Tankard defined a frame as "a central organizing idea for news content that supplies a context and suggests what the issue is through the use of selection, emphasis, exclusion, and elaboration."[37] The events of September 11 were often framed through the visual portrayals. Messaris and Abraham have suggested that visual framing is different than verbal framing because of three distinct properties of visual images—analogical quality, indexicality, and "lack of an explicit prepositional syntax."[38] Analogical quality refers to the idea that the relationship between most visual images and their meanings are based on analogy or similarity to the objects that they represent.[39] Similarly, indexicality suggests that visual images contain direct pointers to objects that give them a "true-to-life" quality that causes people to believe visual images are more accurate than other forms of communication.[40] In the context of framing, this could "diminish the likelihood that viewers would question what they see."[41]

The third distinct property of images is that they lack "explicit prepositional syntax." Unlike the first two properties, this refers to the relationship between images and suggests that the connections are "loose, imprecise and unsystematic," which is opposite of the properties of verbal language.[42] That is, when we use verbal language to communicate, we use certain types of syntactic devices to make propositions or connections, whether we are asserting causality or making generalizations. Since images lack this "explicit prepositional syntax," a viewer's ability to make sense of a series of images is based on other cues. In the context of video, this means that the ways in which individual shots are edited together can convey different kinds of meanings to viewers. One example of this is the phenomenon of associational juxtaposition, an editing device that, in essence, allows the qualities of an object or a person in one image to be transferred to an object or person in the next image.[43]

A case study of the September 11 attacks on CNN shows how the images framed that event. CNN's first fifty minutes of coverage showed only live images of the World Trade Center towers burning and smoking from a variety of angles and distances and both live and recorded images of the second plane flying into the south tower. Between 9:03 a.m. when the second crash occurred and 9:30 a.m., CNN replayed the crash video eleven times. Of those eleven times, the images were shown twice in slow motion with some freeze frames (particularly the frame that showed the impact with the tower). Once the images highlighted the airplane with a digitally added bright spotlight. A journalist did not appear on air until Aaron Brown began broadcasting from a rooftop in New York City with the smoking towers in the background at about 9:35 a.m.

After the first hour of coverage, CNN used two different split screen techniques to show viewers images simultaneously. The term split screen refers to dividing the screen into two "sides" so that viewers can watch more than one image at a time. CNN's initial split screen technique involved showing a smaller box for video on the left and a larger box for video on the right. The trend throughout the first twelve hours of coverage was to use this type of split screen to show a press conference, reporter live shot, or interview "sound bite" in the smaller box on the left side of the screen while watching either live or replayed edited images in the larger box on the right side.

A second standard split screen technique appeared about two hours into the breaking news coverage when CNN was trying to broadcast live from both New York and Washington simultaneously after the Pentagon crash occurred. This second technique would also appear later in coverage when longer, more extensive interviews became a focus of coverage. This second standard split screen technique used two boxes of equal proportion. When used to show two breaking news scenes simultaneously, CNN typically aired "live" video from New York on the left side of the screen and "live" video from Washington on the right. When this equal proportion split screen technique was used for longer interviews, typically the journalist would appear on the left and the interview subject on the right. This equal proportion split screen was usually replaced by the initial split screen as interviews progressed. When this happened the interviewee or "sound bite" appeared in the smaller left box and edited or live video continued to appear in the larger right box.

Throughout the day the images on the right side of the screen were of the plane crashing into the tower (from a variety of angles), the towers smoking and burning, and the towers collapsing. Interspersed among them was shot after shot of terrified and often injured people fleeing the debris clouds from the collapses. CNN also repeatedly aired shots of rescue workers in the haze. These shots were designed to take viewers to what was later called "Ground Zero" to give them a stronger sense of place. These images from New York, which dominated the afternoon coverage, were intermixed with "live" shots as well as edited video from Washington during much of the late morning hours as the Pentagon crash story began to emerge.

Because of the incredible difficulty in managing and producing such an enormous "live" event for television, CNN producers made no apparent effort to connect the information the viewers were hearing from sources on the left side of the screen to the images they were watching on the right. The one exception to this came at 3 p.m., when at the beginning of the hour, an anchor in Atlanta recapped the events of the day to fairly well-timed and matched edited video. This was also the time when afternoon viewers saw images from the Pentagon fire and the Pennsylvania crash site. Otherwise, the general World Trade Center shots mentioned above dominated coverage after about 1 p.m. Given the graphic nature of these images, they effectively showed the devastation and "horror" of the attacks and likely caused a heightened response to what viewers were hearing. Although the images seemed unrelated to the audio in a literal sense, they created a strong sense of urgency and also could have potentially invoked greater fear. CNN almost certainly was not intentionally trying to frighten people, but given the nature of the images, that end result seemed apparent. This magnification effect was even illustrated in some of the comments that CNN journalists made. For example, at one point CNN analyst Jeff Greenfield suggested that the number of fatalities could be near twenty thousand. He said his figure was based on both factual information and his impressions from the images he watched throughout the day.

The video that CNN aired during this first twelve hours was powerful because the reality of the images was so arousing and overwhelming. It is extremely rare to see footage of an airplane crashing into a skyscraper. Add to that footage the effect of slow motion or freeze-frame effects and you get a heightened viewer response.[44] It is not as unusual or arousing to see footage of a building collapse because old buildings are often imploded in front of both news and sometimes movie cameras. But, the context of this footage, coupled with the edited series of shots that often followed the shots of the dramatic collapse, is what was striking here. CNN continuously "looped" or replayed multiple times a series of edited shots that began with a tower collapse, followed by a shot of the massive debris cloud rolling into the streets, people frantically trying to escape, shots of the ensuing darkness, and ending with shots of debris covered and injured people seeking help from emergency workers. This edited series of shots and several others like it took the less personal but dramatic and frightening images of the towers collapsing and personalized them for an even more fear-invoking and arousing effect.

It was in this context that we heard journalists trying to help viewers understand what they were seeing. Even the president acknowledged the incredible impact of the day's images:

CNN Anchor Aaron Brown (After a replay of the second place crash): You know, we
have seen this now, we have seen this honestly dozens of times. And it's no less powerful and no less sickening to see it again. Again, let's just look at this scene. This is
amateur video. The plane coming and now—*no words, no reason* (our emphasis).

CNN Anchor Paula Zahn: As I stand here on this balcony tonight and look back at the
smoke continuing to billow from the wreckage of the towers, it makes you sick.

President George W. Bush: The pictures of airplanes flying into buildings, fires burning, huge structures collapsing have filled us with disbelief, terrible sadness, and a
quiet, unyielding anger.

One additionally powerful and arousing series of images was shown during the first twelve hours of coverage on CNN—Palestinians celebrating in the
streets of East Jerusalem. This video appeared three times between about 3:35
p.m. and 3:45 p.m. Although the images without the context of the terrorist
attacks would not seem to provoke an emotional viewer response, these images juxtaposed with the images of the death and destruction in the United
States were further examples of ways that seemingly disconnected video could
reinforce the general themes offered in the audio and create strong arousal
within viewers. This fits the notion of associational juxtaposition discussed
earlier.

The culminating effect of these images provides additional support for the
three themes that emerged to create the dominant frame in CNN's breaking
news coverage. Seeing Americans coming together to rescue and help each
other showed unity. Seeing the devastation and magnitude of the attacks in
such arousing ways would support an emotional response to the notion that
when one is attacked, one retaliates. Finally, on a different level, seeing the
devastation and "horror" of the attacks gave unspoken justification to the retaliatory response. Perhaps CNN anchor Aaron Brown noted this most strongly
when he said, while looking at a shot of the smoking New York City skyline
in the evening hours, "just take a moment and try and absorb, not with the
facts, not with the pieces of information, but just look at that scene and think
about what happened today. There were 50,000 or so people who came to
work on a beautiful, late summer morning here in New York in those two
towers that are now gone. These are people with families, with children, people who had offices and have them no more, people whose lives are forever
changed."

Conclusion

Images play a key role in the media coverage of terrorism. First, images in general influence memory and perception and have shown potential significant effects. Media organizations have altered images to remind the public of the reasons behind a military intervention. Images help make issues seem real, but that reality is compromised when media organizations use generic images in ways that have furthered stereotypes. Images are an easy and efficient way to draw a reader into a story or to wait through commercial to see the "just in" video. It is unrealistic to assume that the media will not use images of terrorism in their coverage. In fact, showing the awful truth is as important as writing about it. Images will always accompany terrorism stories, providing a significant part of the content that the audience receives.

Notes

1   Kelly McBride, "Did Powerful Image Present an Unbalanced View?" Posted April 10, 2003, http://www.poynter.org/column.asp?id=53&aid=29510.

2   Susan D. Moeller, *Shooting War: Photography and the American Experience of Combat* (New York: Basic Books, 1989).

3   Shahira Fahmy and Daekyung Kim, "Picturing the Iraq War: Constructing Images of Conflict Revisited" (Paper presented at the annual meeting of the International Communication Association, Seoul, Korea, July 2002).

4   Yung-Soo Kim and Zoe Smith, "News Images of the Terrorist Attacks: Framing September 11th and Its Aftermath through the Pictures of the Year International Competition" (Paper presented at the annual meeting of the AEJMC, Kansas City, Mo., August 2003).

5   See, for example, Gavriel Saloman, *Interaction of Media, Cognition, and Learning* (San Francisco: Josey-Bass, 1979); Benjamin H. Detenber and Byron Reeves, "A Bio-informational Theory of Emotion: Motion and Image Size Effects on Viewers," *Journal of Communication*, 46, 3 (1996): 66–84.

6   John E. Newhagen and Byron Reeves, "The Evening's Bad News: Effects of Compelling Negative Television News Images on Memory," *Journal of Communication*, 42 (1992): 25.

7   Examples include http://www.everwonder.com/david/worldofdeath/ and http://www.liveleak.com.

8   Frank Craig, "Why Publish Images of Death," *Pittsburgh Tribune-Review*, Tuesday, June 22, 2004. http://www.pittsburghlive.com/x/pittsburghtrib/s_199849.html.

9   Dan Kennedy, "The Daniel Pearl Video: A Journalist Explains Why Its Horrific Images Should Be Treated as News," *Nieman Reports* (Fall 2002): 80.

10 Show aired on CNN June 19, 2004 at 9 PM; transcript retrieved from http://transcripts.cnn.com/TRANSCRIPTS/0406/19/i_c.00.html.

11 See, for example, W. Russell Neuman, Marion R. Just, and Ann N. Crigler, *Common Knowledge: News and the Construction of Political Meaning* (Chicago: University of Chicago Press, 1992); Barry

Gunter, "Remembering Television News: Effects of Picture Content," *The Journal of General Psychology*, 102 (1980): 127–133; Doris Graber, "Say It with Pictures," *Annals of the American Academy of Political and Social Sciences*, 546 (1996): 85–96.

12 Newhagen and Reeves.

13 Shahira Fahmy, Sooyoung Cho, Wayne Wanta, and Yonghoi Song, "Visual Agenda Setting After 9-11: Individual Emotion, Recall and Concern about Terrorism," *Visual Communication Quarterly*, 13 (2006): 7.

14 Dan G. Drew and Thomas Grimes, "Audio-Visual Redundancy and TV News Recall," *Communication Research*, 14 (1987): 452–461.

15 See the following for a more substantial discussion of the literature on learning from images: Barrie Gunter, *Poor Reception: Misunderstanding and Forgetting Broadcast News* (Hillsdale, N.J.: Lawrence Erlbaum Associates, 1987) and Allan Paivio, *Mental Representation: A Dual Coding Approach* (Oxford: Oxford University Press, 1986).

16 Newhagen and Reeves, 25–41.

17 John E. Newhagen, "TV News Images That Induce Anger, Fear, and Disgust: Effects on Approach-Avoidance and Memory," *Journal of Broadcasting & Electronic Media*, 42 (1998): 265–277.

18 Norbert Mundorf, Dan Drew, Dolf Zillmann, and James Weaver, "Effects of Disturbing News on Recall of Subsequently Presented News," *Communication Research*, 17 (1990): 601.

19 Leonie Huddy, Stanley Feldman, Gallya Lahav, and Charles Taber, "Fear and Terrorism: Psychological Reactions to 9/11," in Pippa Norris, Montague Kern, and Marion Just, eds. *Framing Terrorism: The News Media, the Government, and the Public* (New York: Routledge, 2003), 281–302.

20 Michelle Slone, "Responses to Media Coverage of Terrorism," *Journal of Conflict Resolution*, 44 (2000): 522.

21 Brigitte L. Nacos and Oscar Torres-Reyna, *Fueling Our Fears: Stereotyping, Media Coverage, and Public Opinion of Muslim Americans* (Lanham, Mid.: Rowman & Littlefield, 2006).

22 Brigitte L. Nacos, Yaeli Bloch-Elkon, and Robert Y. Shapiro, "Post-9/11 Terrorism Threats, News Coverage, and Public Perceptions in the United States," *International Journal of Conflict and Violence*, 1 (2007): 105.

23 David S. Broder and Dan Balz, "How Common Ground of 9/11 Gave Way to Partisan Split," *Washington Post*, July 16, 2006: A01 retrieved from http://www.washington-post.com/wpdyn/content/article/2006/07/15/AR2006071500610.html.

24 Shana Kushner Gadarian, "Beyond the Water's Edge: Polarized Reactions in the War on Terror" (Paper presented at the annual meeting of the Midwest Political Science Association, April 12, 2007).

25 Fahmy et al., 4–15.

26 Barry Gunter, Adrian Furnham, and Gillian Gietson, "Memory for the News as Function of the Channel of Communication," *Human Learning*, 3 (1984): 265–271; Doris Graber, *Processing the News: How People Tame the Information Tide* (New York: Longman, 1988).

27 See Allan Paivio, *Mental Representation: A Dual Coding Approach* (Oxford, England: Oxford University Press, 1986).

28 Drew and Grimes.

29 Gadarian.

30 See Steven Livingston, "Clarifying the CNN Effect: An Examination of Media Effects According to Type of Military Intervention." John F. Kennedy School of Government's Joan Shorenstein Center on the Press, Politics and Public Policy at Harvard University, June 1997; Fred H. Cate, "'CNN Effect' Is Not Clear-Cut," *Humanitarian Affairs Review* (Summer 2002), available at http://globalpolicy.org/ngos/aid/2002/summercnn.htm.; Eytan Gilboa, "The CNN Effect: The Search for a Communication Theory of International Relations," *Political Communication*, 22 (2005): 27–44.

31 Robert Fisk, "Remember 'the Whys,'" in David Wallis, ed. *Killed: Great Journalism Too Hot to Print* (New York: Nation Books, 2004), 377.

32 Brigitte Nacos, *Mass-Mediated Terrorism: The Central Role of the Media in Terrorism and Counterterrorism* (Lanham, Mid.: Rowman & Littlefield, 2007), 13.

33 William A. Gamson, "Foreword," in Stephen D. Reese, Oscar H. Gandy, Jr., and August E. Grant, eds. *Framing Public Life: Perspectives on Media and Our Understanding of the Social World* (New Jersey: Lawrence Erlbaum Associates, 2001), ix–xi.

34 Gamson, x.

35 Robert Entman, "Framing: Toward Clarification of a Fractured Paradigm," *Journal of Communication*, 43 (1993): 51–58.

36 Entman, 52.

37 James W. Tankard, Jr., Laura Hendrickson, J. Silberman, K. Bliss, and Salma Ghanem, "Media Frames: Approaches to Conceptualization and Measurement" (Paper presented at the annual meeting of the Association for Education in Journalism and Mass Communication, Boston, Mass., 1991) and James W. Tankard, Jr., "The Empirical Approach to the Study of Media Framing," in Stephen D. Reese, Oscar H. Gandy, Jr., and August E. Grant, eds. *Framing Public Life: Perspectives on Media and Our Understanding of the Social World* (New Jersey: Lawrence Erlbaum Associates, 2001), 95–106.

38 Paul Messaris and Linus Abraham, "The Role of Images in Framing News Stories," in Stephen D. Reese, Oscar H. Gandy, Jr., and August E. Grant, eds. *Framing Public Life: Perspectives on Media and Our Understanding of the Social World* (New Jersey: Lawrence Erlbaum Associates, 2001), 215–226.

39 Messaris and Abraham, 216. For more on this concept see Paul Messaris, *Visual "Literacy": Image, Mind and Reality* (Boulder, Colo: Westview Press, 1994).

40 Messaris and Abraham, 217; Paul Messaris, *Visual Persuasion: The Role of Images in Advertising* (California: Sage, 1997).

41 Messaris and Abraham, 217.

42 Messaris and Abraham, 219.

43 Maria Elizabeth Grabe, "The South African Broadcasting Corporation's Coverage of the 1987 and 1989 Elections: The Matter of Visual Bias," *Journal of Broadcasting and Electronic Media*, 40 (1996): 153–179.

44 Brooke Barnett and Maria Elizabeth Grabe, "The Impact of Slow Motion Video on Viewer Evaluations of Television News Stories," *Visual Communication Quarterly*, 7 (2000): 4–7. This experiment shows that when viewers watch slow motion video they perceive the images to be more intense.

CHAPTER 5
Television and Terrorism

It is quite astonishing and clearly quite deliberate, they intended to put this on television around the world so everybody can see what the people who did this think of America.
> —BBC diplomatic editor Brian Hanrahan, reporting live after the September 11 attacks in the United States

The satellite will distribute terrorist paranoia around the world in living color to match each acceleratingly disruptive event.
> —Marshall McLuhan and Bruce Powers, The Global Village, 1989.

For the terrorist interested in mass coverage, television is still the gold standard, and often terrorists plan their attacks accordingly. Historically, and today, some terrorist attacks are not about conveying a larger message, but rather about instilling fear in those who actually witness the act, disabling a city, or putting a candidate out of an election. For the attacks where media attention is the main goal, television is the medium of choice. The two planes in the New York City terrorist attack on September 11 were timed in such a way to ensure television cameras would be aimed at the first tower as it burned, in time to catch the second plane making impact. And so it was that viewers saw the second plane crash into the second tower and soon heard reports of the attack on the Pentagon and the foiled terrorist attack in Pennsylvania—a powerful succession of images and reports that magnified the horror of the day; it seemed to the audience that these attacks just kept coming. The terrorists chose targets that symbolized military and financial power in the United States, and this well-executed plan played out as expected to vast audiences.

Television was the main stage for this drama. When terrorists attack, people in the United States turn to their televisions for news. Despite talk of new media, old media such as television and radio hold their own when it comes to a portion of the terrorism news audience. Polls show that television is the first place that people turn for news coverage of terrorism, with 91 percent of those surveyed in October of 2001 saying that television news was a useful source of information about the terrorist attacks.[1] Television is also the number one news source generally for people in the United States.[2] A survey for the Council for Excellence in Government showed that television and radio were the highest ranked sources of information for preparing news during the threat of a terrorist attack and in the event an attack occurs.[3] In fact, during the 9/11 attacks, those using Google to search for news were greeted with a screen that told them to turn to their televisions for news.[4]

People in the United States are also watching coverage of international terrorism. In the years since 9/11, Americans have watched television coverage of train bombings in India, Spain, and London. Surveys show that the public is both interested and superficially knowledgeable about terrorism, even when the terrorist act occurs abroad, despite little American interest in general in foreign news. For example, a 2005 Pew Research Center poll showed that 48 percent of people said that they paid close or very close attention to news coverage of the July 7, 2005, London terrorist bombing.[5] This is an extremely large following for a foreign news event. Television news gets serious after an attack, producing more democracy-promoting coverage about serious issues facing the electorate. A study by the Project for Excellence in Journalism two months after the September 11 attacks found an increase in stories about government, the military, national and international affairs, and a decrease in stories about celebrities and lifestyle.[6] But these spikes in coverage and interest do not last for other socially significant stories; they are reserved only for terrorism stories. Terrorism coverage increased 135 percent in the four years after 9/11 compared to the four preceding years.[7]

After a terrorist attack there is a surge in interest in other serious stories on the news. A Pew survey in 2004 found that because of a high level of interest in the war in Iraq, Americans spent more time watching and reading other news, too. This survey also showed a sharp rise in the percentage of Americans who say they closely follow international news most of the time, rather than just when important developments occur, increasing from 37 percent in 2002 to 52 percent in 2004.[8]

Television is an obvious choice to depend on for terrorism coverage because TV news employs immediacy, intimacy, and imagery as core criteria for determining what is broadcast on air. These values make television news the most effective global delivery system for terror events. Easy access is one of the main reasons that people turn to television and radio during breaking news. In the case of September 11, many people needed to see it for themselves before believing that it was real. Television provided the next best thing to a firsthand perspective, with its combination of audio and visual, seemingly real-time tempo, and juxtaposition of video images taken at different times and locations. Scholars have argued that television's dramatization of human emotions makes it qualitatively different than print based news, creating drama in both newsrooms and society for the unfolding melodrama and turning news events into occasions for collective experiences of emotions.[9]

Television News Coverage of Terrorism

Understanding why the audience seeks out television news during crisis is just one part of understanding the role of television in terrorism coverage; the other aspect is examining the content of that news. Terrorism is heavily covered by the news media because it fits the basic definitions of newsworthiness; it is significant, timely, and novel. Even though terrorism is clearly newsworthy, television news has given disproportionate time to terrorism for years, and that has only increased in the years following the September 11 attacks. One scholar noted that in 2006 alone, U.S. national news aired more than seven hundred news stories on terrorism, adding to the more than four thousand three hundred stories related to terrorism that aired in the previous four years. That makes five thousand terrorism stories compared to 138 stories on poverty, 592 segments on education, and 724 stories on crime during the same time period.[10]

Content analysis also shows that terrorism and war do appear to shift the focus on news from domestic to international events. There's been more coverage of foreign policy and global conflict on the network evening news since 9/11, but less coverage of domestic issues, according to ADT Research's Tyndall Report. The number of minutes devoted to coverage of foreign policy was up 102 percent, while coverage of armed conflict rose 69 percent, and coverage of terrorism rose 135 percent. Coverage of crime and law enforcement as well as science and technology dropped by half, and stories on issues involving alcohol, tobacco, and drugs dropped 66 percent.[11]

Even though television devoted more time to terrorism coverage, this in no way suggests the coverage is high quality. Because television is such an extemporaneous medium, live coverage of terrorism is subject to all the foibles of live reporting. Rolling twenty-four-hour news encourages an exponential growth in journalists and other so-called experts who speculate about the nature of potential security threats and likely responses by government and military forces. CNN prime anchor and senior correspondent Judy Woodruff said that the TV pundits "parade as journalists, but have never paid their dues. The concept of accountability is alien; all that matters are attention and ratings."[12] As *New Scientist* news editor Fred Pearce said about his time as a TV pundit, "If I waved my arms in the air and spoke in sound bites, they would call me an expert, write me a cheque, and invite me back one day."[13] Pearce went on to say:

> Too often pundits do not appear to be acting as independent experts, but to be saying things journalists would like to say themselves. When journalists hide behind pundits in this way, rather as ministers hide behind their own boffins, we are in danger of entering a shadow world where science and truth are early casualties. Like a media variant of mad cow disease, the public is left with no means of diagnosing who the safe pundits are, and who are the crackpots.[14]

The time spent on this speculation, graphic images, and opinion erodes the time spent with real experts making sense of the larger issues and context for terrorism. Historically, television has been criticized for not discussing enough news, a function of a thirty-minute nightly news slot with breaks for commercials. Today's criticism is that TV news discusses everything to death. Technology is such that around-the-clock coverage and live reports mean that we can get more information but not necessarily better information. Scholars have shown that when covering events such as terrorist attacks, television news tends to focus on stories about the specific acts often without providing the needed historical, economic, or social context.[15] Specifically, Iyengar found that for news coverage of terrorism, episodic reports outnumbered thematic reports by a ratio of three to one.[16] Terrorism is covered like an event, rather than an ongoing issue.

Sadly, the traditional and now expected fare of light-weight experts does not enhance public understanding of issues.[17] In part, this is because unlike the McLuhan concept of television as a cool medium where people can become involved and participate in an event without becoming too agitated by what they say, television terrorism coverage causes great anxiety and fear among the public. Even McLuhan, shortly before his death, acknowledged that terrorist coverage is an exception to his cool medium theory. He wrote that global satellites will allow terrorists to use the media to spread terror, and acknowledged that this transformed the basic nature of the medium.[18]

The relationship between television and terrorism fundamentally changed on September 11, 2001, not only because of the magnitude of the attacks, but also because of the role played in the coverage by Al Jazeera. Although formed in 1996, Al Jazeera gained international prominence on October 7, 2001, when it broadcast the first declaration of Osama bin Laden on the September 11 attacks, making the Arab satellite channel a principal source of information for Western people. Although conservative media watchdog groups in the United States have called coverage on the network propaganda, Al Jazeera has consistently shown its editorial independence by criticizing Arab governments and covering previously taboo topics such as the role of women in society and religious involvement in politics. This critical coverage gained Al Jazeera credibility, especially as a source of news in the Arab world and as an alternative source for Western media.[19]

A pattern has emerged for the coverage of terror in television, starting with the murder of Israeli athletes at the 1972 Olympics in Germany, and continuing with hostage crises and bombings throughout the 1970s and 1980s. The pattern that emerged was coverage of the attack itself, which includes death or injured tolls, followed by interviews with elite sources speculating about what the U.S. response would be, as well as emotional interviews with family members of attack victims. If Americans died, then images of coffins flying home and memorial services would follow.[20]

The Significance of Breaking News

Closer scrutiny of live coverage of television is necessary to understand the main news fodder of the American public better. Because few terrorism studies empirically focus on television content, looking at the broader field of breaking news is a useful way to parse out research that also applies to terrorism because one common denomination of nearly all terrorism coverage on television is that it starts out as live, breaking news.

Some media critics speculate that American journalists may disregard their codes of ethics when covering breaking news because of the limited amount of time television journalists have to put a story together and get it on the air.[21] Tight deadlines coupled with an absence of hard information often leads to journalists reporting unconfirmed information and rumors, many of which turn out to be wrong.[22]

Essentially, breaking news becomes the organizational routine; the impact of these routines on a journalist's behavior is well documented in the media sociology literature.[23] Shoemaker and Reese define routines as "those patterned, routinized, repeated practices and forms that media workers use to do their jobs ... Routines form the immediate context, both within and through which [journalists] do their jobs."[24] Berkowitz and Limor recently examined the impact of professional confidence on individual journalistic performance. They defined professional confidence as "based primarily on the journalist's experience: the more experienced the journalist, the more professional confidence he or she has."[25] The study determined that a journalist who displayed professional confidence would be more likely to initiate situations with a higher degree of ethical risk. The study concluded that a journalist who displays professional confidence is likely to handle news stories and news sources individually rather than feeling pressured by the news organization to report a story in a certain way.

In a qualitative study of CNN's breaking news coverage of the Oklahoma City bombing, Reynolds suggested that in the absence of traditional journalistic routines, the strength of ideological influence that an individual journalist exerted was greater in a breaking news situation.[26] Reynolds and Barnett found that journalists in a breaking news situation are more likely to report rumors and use anonymous sources.[27] Other ethical models that some journalists follow include social responsibility and an adherence to the value of objectivity. According to Shoemaker and Reese, "objectivity, although a cornerstone of journalistic ideology, is rooted in practical organizational requirements."[28] For many, an accepted definition of objectivity is fairness and balance in decision making, information seeking, and information presentation.[29] Tuchman says that objectivity is a ritual and that its primary purpose is to defend the product, or media content, from critics.[30] Although most journalists realize that objective methods don't provide guidelines for story, source, or fact selection, they still strive to remain personally detached to remain objective. By eliminating the "intent" to add their personal perspective to stories, journalists can reach "evaluative conclusions and state opinions" because they come from objectively gathered facts.[31] Perhaps this is only how it happens in standard coverage. A decidedly different pattern emerges during crisis and breaking news.

In *Mass Media and American Politics*, Graber suggests three different stages of crisis coverage. During stage one the crisis or disaster is announced with a description of what is happening. The journalist will often direct people to places of safety. Graber identifies the retrieval and reporting of accurate information as the biggest challenge in this stage.[32]

In stage two, the media attempt to correct errors from initial reports and put the situation into perspective. Government officials and their critics try to "shape political fallout from the event in ways that support their policy preferences."[33] In stage three, the media attempt to place the crisis in a larger perspective and begin the preparation for coping with the aftermath. In Graber's model, stage one is where breaking news occurs. She notes, "the unrelenting pressure for fresh accounts often tempts media personnel to interview unreliable sources, who may lend a local touch but confuse the situation by reporting unverified or irrelevant information."[34]

Other studies have specifically focused on terrorism events as a means to study crisis. In a study of the national television news coverage of the September 11, 2001, terrorist attacks, government officials, witnesses, and experts were the most frequently used sources.[35] Framing of the coverage varied among networks, but most framed September 11, 2001, as a disaster or looked at the event from a political point of view.[36]

Other studies have looked at the ways that journalists perform their roles generally and how that changes during crisis coverage, both on television and in print. Nossek and Berkowitz suggest that during regular daily reporting, journalists use a professional narrative that is "a balance between core journalistic values and social pressures from their working world," but when a terrorist attack occurs, the journalists switch to a cultural narrative to guide the public in the direction of "the dominant cultural order."[37]

The Nossek and Berkowitz study evaluated U.S. and Israeli newspaper coverage of two terrorist attacks in Israel. The study revealed that when journalists use a professional narrative, they are distributing information in a timely fashion, with accurate sources, by the deadline assigned. When reporting through a cultural narrative, journalists tend to explain the culture's present, past, and the future. The Nossek and Berkowitz study also found that distance between where an event occurs and where the event is covered shapes event coverage. This happens because journalists attempt to "domesticate" foreign news so that viewers are able to link events that they have experienced to the events that are occurring, leading to an ethnocentric focus.

In a separate study, Berkowitz and Limor suggested that when journalists "domesticate" a foreign news event they begin to treat the event as "their own"; consequently, they will focus on their own national loyalties and adhere less to traditional journalistic norms.[38] When a story is not defined as "their own," journalists will continue to follow standard journalistic practices.[39] Some media critics argue that newspaper coverage is more reliable in a breaking news/crisis situation because newspapers aren't as concerned with speed. For example, Stu Bykofsky, *Philadelphia Daily News* columnist, notes that "print reporters gather material, then write, then their work goes through editors' hands."[40] He argues that cable news networks "air stuff they have not properly checked out."

A case study of how one network covered four key terrorism events at home and abroad offers a glimpse at the different ways that terrorism is initially covered in breaking news. This pilot study explores how breaking news exclusively involving terrorism might directly influence content and help shape the role of journalists in breaking news situations. This current study employs the same coding scheme used in a previous study of the September 11 attacks across four networks.[41]

The first five hours of CNN's breaking news coverage of the 1995 Oklahoma City bombing; the September 11, 2001, attacks in New York and Washington, DC; the 2005 London transport attacks; and the 2006 Mumbai, India, train attacks were analyzed for journalistic roles and conventions in a breaking news context.[42] CNN was selected because it was the most popular news source for the September 11, 2001, attacks; a 2001 Pew survey showed that 53 percent of people turned to cable news as their primary source of news about the terrorist attacks. Further, CNN was the only national continual news coverage option during the Oklahoma City bombing.

Coders identified speakers as working journalists, off-duty journalists, nonjournalists, or unknown. The coder then had to determine the role individual speakers assumed based on their comments about the attacks. The role categories included traditional journalist, expert, eyewitness, social commentator, or other.[43]

Another area this study explores is adherence to traditional journalistic routines such as whether speakers reported information that came from rumor, unconfirmed reports, anonymous sources, or how often the speaker made personal references by saying "I" or "me."[44]

Speaker roles were analyzed to determine how often journalists performed a role other than that of the traditional disseminator. London journalists covering the 2005 London subway attacks were found to have spent the most time in roles other than traditional journalists. They spent a total of 26.5 percent of the time as eyewitnesses, experts, social commentators, or in an "other" category.[45] Just slightly below the amount of coverage for London was the coverage of the September 11, 2001, terrorist attacks and the 2006 India train attacks that totaled about 22 percent of time in a role other than traditional journalist. The least amount of time spent outside of the traditional journalist role occurred when covering the 1995 Oklahoma City bombing with about 3 percent.

A traditional journalistic statement is a customary statement that a viewer expects to hear during a news report. Eyewitness statements, expert remarks, and social comments are not traditional accounts expected of journalists. For example,

> September 11, 2001, terrorist attacks, Aaron Brown: Good Lord. There are no words. You can see large pieces of the building falling. You can see the smoke rising. You can see a portion of this—the side of the building now just being covered on the right side as I look at it, covered in smoke. This is just a horrific scene and a horrific moment. (Sample eyewitness statement)

> Oklahoma City Bombing, Alex Perry: And those two groups have been in a loose alliance for many years, providing each other sort of logistical support, money, weapons, and so on. Generally, SIMI [Students Islamic Movement of India] used to be in a background role, setting up safe houses, providing false identifications. And the LeT [Lashkar-e-Toiba] was the one that was actually carrying out its acts. (Sample expert statement)

> September 11, 2001, terrorist attacks, Judy Woodruff: We want to say God bless the souls of those who have lost their lives today or who are dying or are dying as we speak in hospitals and in places where they cannot be reached. I think that even those out there who may not believe that there is a God at a time like this, we all reach out for a higher being and we want to believe that there is someone who can bring us salvation. (Sample social commentary)

One area noted in the coverage that was not covered in the coding sheet was speculation. Often, journalists speculated about a situation, without actually clarifying or providing context for the event. The coders found that a speculative comment did not fit the expert or social commentary categories because these statements were devoid of certain authority and background knowledge but they also did not seem to offer a point of view. Examples include

> 2005 London subway attacks, Richard Falkenrath: Yes, it does look an awful lot like a jihadist group. Maybe not an al Qaeda per se, but someone who shares their ideology and has learned from their method of operating. There are some other possibilities. It could be a faction of the IRA. It's a little unlikely, given what we know. They've typically targeted symbolic targets rather than civilian targets. It could also be a group trying to protest the G8. But I think most likely, it is al Qaeda or affiliate organization.

September 11, 2001, terrorist attacks, Aaron Brown: I am not precisely sure on this, and I want to tell you when I am not precisely sure, but apparently, a plane or helicopter hit part of the Pentagon itself, as you take a look at the pictures there.

The study also explored how often a journalist makes a personal reference using the pronouns "I" or "me." A total of 273 personal statements were made by journalists during all of the attacks studied. When counting these references, the coders excluded times that journalists used the word "I" as a verbal pause. Journalists also tend to refer to statements that they previously made by saying, "As I mentioned earlier ..." These statements were also excluded. Only those instances where journalists were offering personal views were coded. Some examples include

> 2006 India train attacks, Alex Perry, *Time* magazine correspondent: "Well, I don't think we're going to see anybody admit that they carried out this attack."

> 2005 London subway attacks, Oran O'Reilly, journalist: "Oh, definitely rush hour. I was trying to get on the Tube myself, and I wasn't able to get through."

The content analysis also identified when journalists gave information about unconfirmed reports or rumors or used anonymous sources. There were a total of thirty-six rumors reported in all four events. London journalists reported twelve rumors, Oklahoma City reporters mentioned eleven, and 9/11 reporters, ten. The least number of rumors were found in the India coverage, with only three. Some of the rumors reported included who was responsible for the events, what actually occurred during the terrorist attack, where the events occurred, and what caused the attack. It is important to note that journalists were airing unconfirmed reports, some of which were later confirmed. This happened in Oklahoma City, September 11, and London coverage.

Here's an example of an unconfirmed report during the 2005 London subway attacks from CNN Correspondent Fionnuala Sweeney: "There are reports, as you can imagine, coming in by the second. We're getting reports of even a second blast being heard in London's Tavistock Square. We cannot confirm that independently as yet. And I think it's important to try to differentiate between what we know to be fact and also speculation."

An example of a reported rumor that was later confirmed to be false occurred during the Oklahoma City bombing from a KWTV male anchor who noted, "Two of the men involved, perhaps, are Middle Eastern men. One is twenty to twenty-five. The other is thirty-five to thirty-eight. Both with dark hair and a beard, and they were both wearing blue pants, black shirts, and coats. The driver—[we] don't have a description on him. But we do know a suspect vehicle. And there it is. A brown Chevy pickup with tinted windows. And it had a bug shield on front. It was last seen northbound on Walker. That would be going away from this building, devastated this morning." Of course, Timothy McVeigh, a white U.S. citizen who drove a rental moving truck to the scene, was later convicted and sentenced to death for the bombing.

Journalists attributed a total of only four statements to anonymous sources with three of those coming from journalists covering the Oklahoma City bombing and the other from a journalist reporting on the London subway bombing. A KFOR male anchor attributed information to an anonymous source while covering the Oklahoma City bombing. He said:

> We did receive a phone call earlier this morning, shortly after this occurred, from somebody claiming responsibility … But somebody called News Channel 4 and tried to claim responsibility, saying that they were with the Nation of Islam. We have not been able to confirm that report. We are still working on that. We certainly do not want to do a disservice to the Islamic community. So please take that with the sprit that it is intended, and that is that that is an unconfirmed phone call that we received earlier this morning.

This study also looked to see if these journalistic conventions varied by event and found some statistically significant differences. Journalists reporting on London, Oklahoma City, and September 11 were all more likely to report rumors than those who covered the India bombings. Journalists covering Oklahoma City and London were more likely to make anonymous attributions and journalists covering September 11 and London were more likely to make personal references.

This content analysis also examined if the role of the journalist varied by event. As mentioned, the most time that each reporter spent with each event was as a traditional journalist. During the Oklahoma City coverage, journalists spent almost 97 percent of the time in the role of a traditional disseminator. About 78 percent of the time was spent as traditional journalists during the September 11 and India coverage, and about 73.5 percent during London coverage.

Eyewitness accounts were the highest for September 11 at about 18.5 percent. The journalists covering the India train attacks spent almost 11 percent of the time in the role of an eyewitness, where London journalists spent about 8.5 percent. The fewest number of journalists acting as eyewitnesses was during coverage of the Oklahoma City bombing at less than 1 percent.

There was also a difference in the amount of expert coverage each event received, and this breaks down between the two international events and the two domestic events. The most expert coverage came from the two international events where London had almost 16 percent of coverage time devoted to expert analysis, while coverage of India featured expert commentary almost 8 percent of the time. The two domestic events had the least amount of expert coverage with September 11 offering about 2 percent expert coverage and Oklahoma City less than 1 percent.

The two international events saw about 2 percent of the coverage time spent with journalists acting as social commentators, and the domestic events about 1 percent.

Time Journalists Spent in Each Role by Event

	Oklahoma City Bombing	9/11 Attacks	London Bombing	India Train Attacks
Traditional Journalist	96.75 %	78 %	73.53 %	78.41 %
Eyewitness	0.94 %	18.52 %	8.54 %	10.71 %
Expert	0.85 %	2.23 %	15.81 %	7.59 %
Social Commentator	1.46 %	1.15 %	2.07 %	2.24 %
Other	0	0.10 %	0.05 %	1.05 %

Substantial differences were also found in the amount of time spent covering the breaking news event during the first five hours of coverage. The two domestic events—Oklahoma City and September 11—were covered exclusively in the first five hours. CNN did not break away for any other stories. In its coverage of the London and India attacks, CNN spent time dedicated to other topics (9.1 percent for London and 45.6 percent for India). This data suggests that American news focuses exclusively on the terrorist event when the coverage is directly related to the United States and covering several other stories while reporting on topics that are not as closely related.

Two main observations may be derived from this pilot study. It appears that violations in journalistic norms during terrorism coverage are not decreasing. There was no apparent pattern in the number of unconfirmed reports or rumors decreasing as the years went on. It would make sense for CNN journalists to learn from their mistakes and to report fewer rumors and unconfirmed reports as time progressed. However, this was not the case. Even after the critical lens placed on CNN after the Okalahoma City bombing, journalists continued to make personal statements and offer social commentary. But, one notable change comes from a qualitative analysis of the transcripts. Whereas journalists in the Oklahoma coverage did not talk about the violations of norms, journalists in the events that followed acknowledged these violations. For example, in the September 11 coverage there was this exchange:

> Jeff Greenfield: We should also mention, I think, Aaron, that inevitably, some of these early fragmentary reports are going to be needing correction, and that will be done as soon as possible.

> Aaron Brown: I think in fairness, there is in a number of places right now—perhaps four or five—chaos and numbers that come out are not necessarily going to hold up, and in our reporting, we will be a bit conservative on some of this until we track it down. There is no point in allowing this thing to seem worse than it is. It is already horrendous, and we don't need to make it worse by misstating numbers and we want you to keep that in mind.

Although the quantitative data show that instances of violations of journalistic norms and conventions did not decrease, the qualitative analysis showed that journalists were more aware of these violations and pointed them out to viewers.

Also, this pilot study determined that the way a story is reported could be based on the relationship that the country affected has with the news organization reporting the event. This limited analysis shows that homeland events are covered in more detail than international events, possibly because journalists regard homeland events as more relevant to viewers. And, the amount of time spent covering stories unrelated to the breaking news event indicates that CNN gives preferential coverage to events most closely associated to homeland events. In this comparative content analysis, results showed that decisions about newsworthiness revolved around proximity to the event. This result confirms what has been documented in other studies, particularly the work of

Nossek and Berkowitz—that the coverage of terrorism is most extensive when the terrorism is on American soil, followed by international events in locations strongly connected with the United States, and least covered when it is international terrorism in places with little U.S. interest.

In some ways, this pilot study is limited in what it can conclude. Only CNN was used for analysis, and the number of speaking turns varied between events, which made it difficult to use inferential statistics based on means. As a result, this analysis only gives us a preliminary glimpse of coverage of breaking news related to terrorism events. Still, the findings offer us a better understanding of the ways in which television presents its initial coverage of terrorism. As previously noted, it is breaking news of terrorism on television that reaches the largest audiences around the globe. Several observations made during this study are also worth noting here.

First, given that Britain is a common terrorist target and its media familiar with covering terrorist events, it is interesting to note how the British system of media shaped the coverage of terrorism events in London. A cursory glance at the BBC coverage of the London attacks showed that few outside sources or experts were used. Instead, the BBC employed a vast number of journalists who cover beats and have become experts in their chosen area. The British social responsibility model of journalism might also account for some of the differences that emerged in the coverage. The quantitative measure used here was not well designed to deal with international journalists and cultural difference, but the case study reported in the next chapter—media coverage of terrorism at home and abroad—will address this issue in greater detail.

Finally, the role of other media is worthy of brief discussion. While many point to new media as the best hope for rekindling interest in news, only 11 percent of eighteen-to-twenty-four-year-olds list news as a major reason for logging on to Internet. One of the biggest issues discussed in journalism and journalism education is how the Web changes the way the news reaches people. Specific to terrorism, what are people looking for when they seek out terrorism news online? The role of bloggers has changed the way that people receive information online, but it is still unclear for how many people and in what ways. Bloggers are not a key source of information for breaking news. It's more likely that the audience would turn to the blogosphere for analysis of the media coverage, using sources such as the Tyndall Report.

The Web does play a key role in some graphic aspects of terrorism coverage. As discussed in the previous chapter, an example is video of the execution of *Wall Street Journal* reporter Daniel Pearl. The *Boston Phoenix* posted the video on its Web site, which was declared by critics as carefully crafted terrorist propaganda that is insensitive to Pearl's family. Others pointed out that journalists often make choices to publish material that hurts people. Some journalists defended the decision, stating that these images should be treated as news because they portray the horror and racism of one terrorist group. Even if news sites did not publish the video, people who wanted to see the video could find it. The Web site Orish.com posted the whole video, along with other gruesome photos of autopsies and accidents. Some journalists argue that at least news sites write about the event and try to put it in a context, rather than simply posting the video as part of a site that focuses on graphic, disturbing images.

Conclusion

In 1972, 46 percent of college-age Americans read a newspaper every day. Today it's only 21 percent, according to research by the Roper Center for Public Opinion Research's General Social Survey. Further, today a record number of people find their information about the world from nontraditional news sources such as comedy news shows, entertainment news magazines, and talk shows. The different ways that these shows approach the coverage indicates how young people, for example, may be informed about terrorism. Perhaps soft news outlets such as entertainment news magazines and talk shows have captured viewers who would not normally follow foreign crises. These shows transform political issues into entertainment and thus inform a segment of the population that would not normally be made aware. However, the coverage is likely not doing much to inform the audience. These entertainment type shows tend to focus on the individual rather than looking at the broad political context in which terrorism occurs. The coverage tends to be episodic, following major events such as the Atlanta Olympic bombing, profiling suspected terrorists and examining potential weapons. One entertainment show worthy of examination is *The Daily Show* on Comedy Central, which is setting the agenda for the heavily coveted eighteen-thirty-four demographic. A mere look at the guest line up shows that this show is seen as a place to discuss serious matters such as politics and the war on terror. In fact, the show won the prestigious Peabody award for its election coverage. A seminal study of the coverage in *The Daily Show* showed that the fake news shows were just as substantive as network coverage.[46] However, even entertainment shows also demonstrated deference to the government after 9/11, as late night jokes quickly eased up on President George Bush because it was seen as tasteless to make fun of the commander in chief during a time of crisis.

The role of late night comedy and fake news shows is an area that needs more exploration. With an overall shrinking news audience and a median age of sixty for CNN and network news, more and more of the public, particularly the young public, will be getting their information about the world around them from Jon Stewart, Stephen Colbert, and other entertainers. As author David Mindich has effectively shown, young people are not attracted to traditional news, and their knowledge about terrorism depends on serious news seeping into entertainment programs.[47] With younger generations tuning out, the democratic implications are dire and move the United States away from a better understanding of the root political causes of terrorism.

Notes

1   Guido H. Stempel III and Thomas Hargrove, "From an Academic: Newspapers Played Major Role in Terrorism Coverage," *Newspaper Research Journal*, 24, 1 (Winter 2003): 55–57.

2   Project for Excellence in Journalism 2007 State of the News Media Annual Report on American Journalism http://www.stateofthenewsmedia.org/2007/printable_localtv_publicattitudes.asp.

3   Hart Teeter, "From the Home Front to the Front Lines: America Speaks Out about Homeland Security" (March 2004). http://209.85.207.104/search?q=cache:T0EQjCrTkCoJ:www.excelgov.org/admin/FormManager/filesuploading/Homeland_Executive_Summary.pdf+Council+for+Excellence+in+Government+terrorism+coverage&hl=en&ct=clnk&cd=1&gl=us&client=firefox-a.

4   Brooke Barnett, *The War on Terror and the Wars in Iraq*, Greenwood Library of American War Reporting, 8 vols., ed. David A. Copeland (Westport, Conn.: Greenwood Press, 2005).

5   Pew Research Center Survey, "Republicans Uncertain on Rove Resignation: Plurality Favors Centrist Court Nominee," July 19, 2005. http://people-press.org/reports/display.php3?ReportID=250.

6   Project for Excellence in Journalism, "Before and After: How the War on Terrorism Has Changed the News Agenda," November 19, 2001 http://www.journalism.org/node/289.

7   Project for Excellence in Journalism, "How 9-11 Changed the Evening News."

8   Pew Research Center for the People and the Press, "News Audiences Increasingly Politicized, "Online News Audience Larger, More Diverse," Released June 8, 2004 http://people-press.org/reports/display.php3?ReportID=215.

9   David L. Altheide and Robert P. Snow, *Mediaworlds Postjournalism* (Berlin: Mounton de Gruyter, 1991); David Dayan and Elihu Katz, *Media Events: The Live Broadcasting of History* (Cambridge, Mass.: Harvard University Press, 1992).

10 Shana Kushner Gadarian, "Beyond the Water's Edge: Polarized Reactions to Images in the War on Terror" (Paper presented at the 2007 annual meeting of the Midwest Political Sci-

ence Association. Chicago, Ill. April 12–15, 2007).

11 See http://tyndallreport.com/.

12 Doug Gavel, "CNN's Woodruff assesses TV News: Veteran Journalist Gives Television High Grades for Crisis Coverage," *Harvard University Gazette*, November 8, 2001, http://www.hno.harvard.edu/gazette/2001/11.08/10-woodruff.html.

13 Fred Pearce, "Forum: As Seen on Television/Where Are His Experts Coming From," *New Scientist*, June, 16 1990. http://www.newscientist.com/article/mg12617215.000-forum-as-seen-on-television-where-are-his-experts-coming-from-.html.

14 Pearce.

15 David L. Altheide, *Creating Fear: News and the Construction of Crisis* (New York: Aldine de Gruyter, 2002); David L. Paletz, John Z. Ayanian, and Peter A. Fozzard, "Terrorism on TV news: The IRA, the FALN and the Red Brigades," in William C. Adams, ed. *Television Coverage of International Affairs* (Norwood, N.J.: Ablex, 1982), 143–166.

16 Shanto Iyengar, *Is Anyone Responsible? How Television Frames Political Issues* (Chicago: University of Chicago Press, 1991).

17 Lawrence D. Soley, "Pundits in Print: 'Experts' and Their Use in Newspaper Stories," *Newspaper Research Journal*, 15 (Spring 1994): 65–75.

18 Marshall McLuhan and Bruce R. Powers, *The Global Village: Transformations in World Life and Media in the 21st Century* (Oxford: Oxford University Press, 1989).

19 For an excellent analysis of Al Jazeera see Mohamed Zayani and Sofiane Sahraoui's book *The Culture of Al Jazeera: Inside an Arab Media Giant* (West Jefferson, N.C.: McFarland, 2007).

20 Jeffrey D. Simon, *The Terrorist Trap: America's experience with Terrorism* (Bloomington, Ind.: Indiana University Press, 2001).

21 David Stone and David Hartley, *Media Ethics* (Princeton, N.J.: Films for the Humanities & Sciences, 1998).

22 Amy Reynolds and Brooke Barnett, "'America under Attack' CNN's Visual and Verbal Framing of September 11," in Steven Chermak, Frank Bailey, and Michelle Brown, eds. *Media Representations of September 11th* (Westport. Conn.: Praeger, 2003), 85–101.

23 Pam Shoemaker and Stephen D. Reese, *Mediating the Message: Theories of Influence on Mass Media Content* (New York: Longman, 1996); Stephen D. Reese, "Understanding the Global Journalist: A Hierarchy-of-Influences Approach," *Journalism Studies*, 2 (2001): 173–187.

24 Shoemaker and Reese, 105.

25 Dan Berkowitz and Yehiel Limor, "A Cross-cultural Look at Serving the Public Interest: American and Israeli Journalists Consider Ethical Scenarios," *Journalism: Theory, Practice & Criticism*, 5, 2 (2004): 159–181.

26 Amy Reynolds, "How 'Live' Television Coverage Affects Content: A Proposed Model of Influence and Effects" (Paper presented to the International Communication Association Conference Mass Communication Division, Montreal, Quebec, May 23, 1997).

27 Reynolds and Barnett.

28 Shoemaker and Reese, 112.

29 Dan Drew, "Roles and Decisions of Three Television Beat Reporters," *Journal of Broadcasting*, 16 (1972): 165–173.

30 Gaye Tuchman, "Objectivity as Strategic Ritual: An Examination of Newsmen's Notions of Objectivity," *American Journal of Sociology*, 77 (1977): 660–679.

31 Herbert J Gans, *Deciding What's News—A Study of CBS Evening News, NBC Nightly News* (New York: Random House, 1979).

32 Doris A. Graber, *Mass Media and American Politics* (7th ed.) (Washington, D.C.: CQ Press, 2006).

33 Graber, 133.

34 Graber, 132.

35 Xigen Li and Ralph Izard, "9/11 Attack Coverage Reveals Similarities, Differences," *Newspaper Research Journal*, 24, 1 (Winter 2003): 204–219; Amy Reynolds and Brooke Barnett, "This Just in … How National TV News Handled the Breaking Live Coverage of September 11th," *Journalism & Mass Communication Quarterly*, 80 (2003): 689–703.

36 Li and Izard.

37 Hillel Nossek and Daniel Berkowitz, "Telling 'Our' Story through News of Terrorism: Mythical Newswork as Journalistic Practice in Crisis," *Journalism Studies*, 7, 5 (2006): 691.

38 Berkowitz and Limor, 159–181.

39 Berkowitz and Limor.

40 Stu Bykofsky, "Cable News Too Fast, Not Final," *Philadelphia Daily News*, October 10, 2006 from http://go.philly.com/byko.

41 Reynolds and Barnett, "This Just In."

42 These events were chosen because they are the most recent and significant terrorist attacks.

43 The traditional journalist category was used to describe an anchor or reporter asking an interview question or for correspondents, reporters, anchors, or other journalists who were reporting objective, factual information. The eyewitness category was used to describe someone who was explaining what happened from his or her personal point of view. This did not include commentary or speculative remarks, but did include personal stories about directly related experience or descriptive commentary about what a person witnessed. The expert category was used to describe someone who was trying to explain, clarify, or provide context to the event. The content for the expert category is focused on providing explanations and information that the general public doesn't have. Expert reports were found in the coverage of all four events. The category of social commentary was used to describe someone who was offering his or her point of view about the events; these statements were pure opinion, devoid of an expert's factual context.

44 A coder reliability test was performed on 10 percent of the sample, with 81 percent agreement across all categories. The objective categories had 98 percent reliability, while the subjective categories were lower. The lowest reliability measure was for the social commentator role at 77 percent. The coders analyzed 1,521 total speaker turns from all four of the events. First, the coders considered that journalistic status of each speaker, coding them as "working journalist," "off-duty journalist," "nonjournalist," or "unknown status." The nonjournalists and the journalists categorized under unknown status were then removed from the sample giving 1,041 total journalist speaker turns to evaluate.

45 Journalists spent more time on air during the international events, than the domestics ones.

46 Julia R. Fox, Glory Koloen, and Volkan Sahin, "No Joke: A Comparison of Substance in the Daily Show with Jon Stewart and Broadcast Network Television Coverage of the 2004 Presidential Election Campaign," *Journal of Broadcasting & Electronic Media*, 51, 2 (2007): 213.

47 David T.Z. Mindich, *Tuned Out: Why Americans Under 40 Don't Follow the News* (New York: Oxford University Press, 2004).

CHAPTER 6

Media Coverage of Terrorism at Home and Abroad

The journalists were people living in their own country that had been attacked and there was not enough questioning [about Iraq].

—Christiane Amanpour, 2007 interview

Journalists need to be no less critical, cynical, and challenging during standard political coverage than when the nation is threatened. But, journalists succumb to the same human tendencies as others. Journalists get defensive. They act nationalistically. They are afraid. All this, particularly during crisis, can show in the coverage. The differences in coverage of the same event from one country to the next reflect the culture of each country, the proximity to the event, and the way the audience expectations influence the coverage. Some of the alarmist tone in much terrorism coverage appears to be specific to U.S. culture.[1]

Critics of the BBC's handling of the July 7, 2005, London Underground bombings focused on the restraint the network displayed, cautiously reporting facts only after checking firsthand to see if they were accurate. The attack was the first big breaking news story since the BBC published its new editorial guidelines, which emphasized that "accuracy is more important than speed."[2] Further, previous studies have found that the media have done a good job toning down the intensity of their characterizations of terrorists.[3] For example, a pre-9/11 study of newspaper terrorism coverage showed that journalists' statements tended to be less judgmental and inflammatory than statements made by government officials.[4] Other scholars have explored the business model and how audience demands contribute to poor terrorism coverage, suggesting that "as serious news organizations move increasingly away from reporting what journalists/gatekeepers deem important for the enlightenment of fellow citizens to what profit-oriented corporate managers consider interesting for the entertainment of news consumers, "hard news" is increasingly crowded by "soft news.""[5] The development of twenty-four-hour news and infotainment has helped to proliferate this trend.[6] In the weeks after September 11, when the U.S. media obsessively reported the possibility that terrorists have or could have had access to chemical, biological, and nuclear weapons, the press "dwelled endlessly on the outburst of patriotism and the idea of national unity without paying attention to other important matters in the political realm." They also neglected their important role as a political watchdog.[7]

Critics and scholars have also looked at how media coverage of terrorism suggests fear and panic. For example, media coverage of anthrax hoaxes that followed the September 11 attacks found that the reporting was "characterized by fear rhetoric, speculation, and confusing incidents that contributed to outrage."[8] Media critics have suggested that in order to best uphold the role of political watchdog when reporting on terrorism, journalists should remember that they must choose their words carefully to "avoid the danger of turning terrorism into theater and news into dramatic entertainment."[9] Critics worry that by exploiting terrorism as infotainment, the media effectively become a terrorist accomplice, "merchants of fear," whose inflammatory coverage merely promotes the terrorist message.[10] A study of some of the mistakes journalists have made in reporting on past terror attacks suggested that media organizations develop a set of guidelines for covering terrorism, including refraining from sensational and panicky headlines, inflammatory catchwords, and speculations about the terrorist plans and government response.[11]

Domestic versus International Coverage

The fear-mongering language is one piece of how a terrorist event is framed. Another aspect of framing depends on the proximity of the event to the country reporting on it.[12] The country's historical background in dealing with terrorism can also affect the quality and amount of terrorism news coverage. When a country has a great deal of experience with domestic terrorism, the media might have more perspective than in those countries that are rarely attacked, and thus the event can be framed in an entirely different manner.[13] Whether the event being covered took place internationally or domestically can also affect how the information is portrayed, as evidenced by the extensive coverage in the United Kingdom and the United States of the London 2005 transport bombings as compared to Egypt's deadliest attack in a decade, which occurred two weeks after the London attacks. The attacks in Egypt killed eighty-eight people, compared to fifty-six in London. The London attacks garnered far more media attention and continued to dominate as a news story two weeks later on the day of the Egyptian attacks.[14]

Although Al Jazeera and Egyptian television prominently featured the Egyptian attacks, world news just reported on the new attacks in Egypt but continued to focus in-depth coverage on the attacks in London. This is not surprising since proximity is a key element in determining whether a news story makes it on the air or in print, and whether the public will be interested in it. According to Einar Östgaard, "News concerning persons, things, or issues with which those handling and those receiving news are most familiar finds its way through the news channels more easily than news concerning unfamiliar persons, things, or issues."[15]

There is a natural link between news proximity and reporting in that national news programs are likely to talk about events that affect citizens of their own countries. However, news in the aftermath of a terrorist attack needs to be understood through the lens of nationalistic perception. When citizens are threatened by actual terrorists or the fear of future attacks, calls to consolidate around a national identity are quite common. Further, other studies have shown that journalists report more extensively on events and issues that directly affect them. For example, a case study of the war in Afghanistan found that possible civilian deaths as a result of the war were given the least amount of news coverage because the victims were not American.[16]

Studies have also shown that coverage can vary depending on the country reporting on it. For example, a framing analysis study of both American and Chinese newspaper articles about SARS in China found that the degree and type of frames employed were different depending on whether the coverage was American or Chinese.[17] A study of British coverage of IRA bombings found that when the bombings by Irish terrorists resulted in death or injury to British military or civilians, political motives for the bombings were not discussed, and only British rather than Irish officials were interviewed for television news reports.[18] A framing study of the coverage of the United States' shooting down of an Iranian airplane and Soviets' shooting down of a Korean jet found that "by de-emphasizing the agency and the victims and by the choice of graphics and adjectives, the news stories about the U.S. downing of an Iranian plane called it a technical problem while the Soviet downing of a Korean jet was portrayed as a moral outrage."[19]

Previous content analyses of American terrorism coverage have highlighted the media's tendency to frame acts of terrorism in a one-sided manner, eliminating the need for debate about important issues.[20] One analysis of September 11 news coverage found that the United States was portrayed as a symbol of a morally powerful victim, "ensnared in a position that required it to transform victimization into heroic retributive action," creating a national

identity that made political debate over state action unnecessary and even immoral.[21] A study of CNN's verbal and visual framing of September 11 concluded that media coverage can even suggest governmental retaliatory action,
framing the events as an act of war that justified and required a military retaliation response.[22] A comparative study of television news coverage in the
United States and Britain of a terrorist event in London's Hyde Park in 1982
found that the news reports varied according to perspectives and news practices, emphasizing different visual and thematic aspects of the event.[23] A study
that focused on the war on terrorism in the ten largest newspapers in the
United States showed that "Editorial writers drew selectively on historical references, government sources, and contextual statements in similar ways to
frame the tragedy and the potential U.S. response to it. No editorial suggested
that military intervention would be inappropriate and none stated that military
intervention would not ultimately succeed, although some urged caution."[24]

Further anecdotal evidence from journalists highlights their perceptions
about differences in coverage between the United States and other countries.
Independent journalist Dahr Jamail argues that the events he covers are often
unpalatable for an American audience because they detail the harsh truth of
terrorism:

> My potential audience has no power or sway over what I write. I write about events
> that are happening in real time, in Iraq and around the Middle East. If this is upsetting
> to a certain audience, then they can watch CNN or Fox news instead.

> This of course limits my audience in the U.S. to alternative media outlets, and tends
> to find me a far larger audience abroad. While I am regularly on national T.V. and ra-
> dio outlets in European and Asian countries, in the U.S. my audience tends to be lim-
> ited to alternative media outlets.[25]

However, a cross-cultural study of 2003 Iraq War coverage found that
even though news stories in Sweden and the United States were for the most
part neutral, the Swedish newspapers were more negative. This possibly suggests that the audience expectation had an impact, because during the time of
the study the majority of people in Sweden opposed the war.[26] Further, a
comparative analysis of U.S. and Arabic news showed that Al Jazeera and U.S.
networks, except for Fox News Network, were balanced. In contrast, 38 percent of Fox News Network stories were supportive of the war in tone.[27]

Finally, there is a common assumption in Europe that American news is not very sophisticated. Peter Jennings was asked about this in a live forum with the BBC three months after the September 11 attacks and he responded:

> Well I just think it is inaccurate. I think that people in Europe tend to believe that because America does not pay as much attention to the rest of the world as the rest of the world pays to America that our news is dumbed down. One of the things that needs to be essentially understood about the United States is that we have more information available to the public than I think any other nation on earth, whether it's on television, radio, in our vast number of newspapers and magazines—opinion of every imaginable position can be read and seen and absorbed here. I don't think that television always does the absolute best, most sophisticated job of covering the world—we would like to have more time to do so. But I think that to suggest that the news in America is dumbed down is somewhat ill-informed.

British and American Terrorism Coverage

To expand upon the work of previous studies of media and international terrorism, this chapter qualitatively compares mainstream British and American television and newspaper terrorism coverage to determine if that coverage follows the dominate hegemony as well as differences in coverage when a terrorist attack occurs domestically and internationally and differences in content between the American and British coverage. The analysis of both broadcast and print media will provide for both the day of and day after initial coverage of the event, and also the in-depth context stories in newspapers that come in the weeks that follow an attack. By choosing the breaking news segments for television as well as continual newspaper coverage, a broad news cycle of the terrorist event will be covered.

This case study is grounded in framing theory. As noted in earlier chapters, Entman suggests that "to frame is to select some aspects of a perceived reality and make them more salient in a communicating text, in such a way as to promote a particular problem, definition, causal interpretation, moral evaluation, and/or treatment recommendation for the item described."[28] News frames are tools that media have developed to interpret and convey information.[29] As a result of journalists selecting which parts of the story to tell and which elements to exclude, news events can be "framed in various ways, producing different versions and different attributes,"[30] thereby increasing the salience, importance of the event, or certain elements of the event. Frame analysis, however, is not an audience study and can only suggest how an audience might receive the framed message.

Comparing international and domestic coverage also provides insights into the differences in the press of the United States and abroad. The U.S. mainstream press is not known for extensive coverage of international issues. U.S. newspaper editors and television news executives have reduced the space and time devoted to foreign news coverage by 70–80 percent in the past two decades. For some time now, content studies have documented the lack of international news coverage in the United States, and public opinion polls have shown the lack of interest in international news.[31]

Only a small number or people in the United States claim any interest in non-war zone coverage of international events. So when the U.S. press does cover the world, it covers the hot spots of terrorism, war, and natural disasters. One study showed that nine out of ten international stories on the local news were about war, violence, or disaster. And even in these stories, the sources are rarely international, but rather focused on U.S. military or government response to the international conflict. Terrorism coverage constitutes the small bit that we learn about the outside world. But policy coverage is rarely the focus, so the political context of terrorism is not part of the news cycle.[32] An analysis of the Pew Center Study Public Attentiveness to New Stories from 1986 to 2006 reveals that domestic news stories dominate interest. During this twenty-year period, only three years showed foreign news stories receiving top interest. They were the death or Princess Diana in 1997, the news from Iraq in 2003, and the Iraq invasion of Kuwait in 1990. Arguably the two war topics are not really about interest in foreign news stories but rather interest in a U.S. story that takes place on foreign soil.[33]

In this chapter, comparative analysis examines both broadcast and print terrorism coverage of the September 11, 2001, terrorist attacks in New York and Washington, D.C., and the July 7, 2005, London Underground and bus bombings. The United States's CNN and the United Kingdom's BBC are the focus of the analysis. CNN was selected for this analysis because U.S. viewers consistently considered it their primary news outlet during the September 11 terrorist attacks, according to Pew Research Center polls. In addition, previous studies of breaking news coverage of September 11 showed strong similarities in coverage on ABC, CBS, NBC, and CNN.[34] Ratings data from the United Kingdom show that the BBC is the primary source of news in the United Kingdom during times of crisis, and that during the July 7, 2005, bombing, the BBC drew the largest audience.

Four newspapers—the *New York Times*, *USA Today*, the *London Times*, and *The Guardian* of London—were chosen because they represent high circulation newspapers and are the most well-regarded newspapers in each country. *USA Today* has the highest American newspaper circulation. The *New York Times* ranks third in circulation and is also widely considered the paper of record in the United States. The British *Guardian* and the *London Times* were also selected because of their relatively high circulation. In addition, the *Times* has been widely regarded as Britain's newspaper of record. Newspapers were included in this study to illustrate how the differences between the British and American press were manifested in their terrorism coverage. Studies have shown that the British press employs a more interpretive style of writing than is typical in North American papers, and British papers are also more likely to be divided along party lines than American newspapers.[35]

The following research questions were used:

RQ1.   How did British and U.S. television and newspapers frame the 9/11 attacks?

RQ2.   How did British and U.S. television and newspapers frame the 7/7 attacks?

RQ3   How did coverage vary by event and also country reporting?

Let's discuss the first research question now: all of the coverage of the September 11 attacks from the United States and Britain emphasized the gravity and shock of the attacks. The U.S. coverage focused mainly on the seriousness and surprise and the British coverage emphasized the surprise at first, but quickly moved on to put the event in historical context, comparing the attacks to other acts of terrorism against the United States, such as the 1993 World Trade Center attacks, but also the Oklahoma City bombing and the 1996 Atlanta Olympic Park bombing.

At first, the British coverage tended to emphasize that the United States did not have a great deal of experience with domestic terrorism and had no inclination of a potential attack. The attacks were said to have illustrated the United States' vulnerability. Statements such as these were typical:

The Guardian Sept. 12: The United States is unlikely ever to be the same again in the wake of this onslaught. The country was hit, with great deliberation, at the very core of its economic and military power, presumably a message to Americans that they will never be able to consider themselves safe.

BBC correspondent in New York City: "I think the atmosphere is one of absolute surprise. People are just stunned. Nobody could imagine this in one of the most densely populated cities in the world.

The British media covering September 11 also tended to give conflicting reports about how the people on the scene were handling the crisis. On the one hand, the media reported chaos and devastation, but they also, at times, portrayed a TV image of surprising order and calm. Statements that reflected both the chaos and panic of the situation and the relative calm of the victims were fairly common. U.S. journalists often showed fear and calm in the same scenes. CNN used a split screen of video and a crawl that was often not semantically linked. Viewers were seeing taped earlier footage of people running at ground zero while the buildings collapsed at the same time the Mayor of New York was talking about how calmly people were walking out of the city. The calm walking across the bridge footage was sometimes shown, but at other times it was a shot of the mayor, leaving viewers to think that he was talking about how well the city was handling the crisis as they watched chaos on the second half of the screen. A scene such as this would be followed with the suggestion that more attacks were likely to come. Often reporters corrected themselves in midsentence as they suggested more chaos than was present:

BBC reporter: at which point people came—I nearly said screaming; they weren't screaming, they were—it was a mild panic, if you like. People were simply saying, get out of here, get out of here ... we all streamed out, some people running, some people crying, nobody really screaming ... The cause, I cannot illuminate. All I can tell you, it was a very frightening experience, but people by and large reacted very well to it.

The British media also tended to place more of an emphasis on reporting how the crisis was being handled by the government and emergency services, and presenting information for the benefit of the public:

The Guardian Sept. 13: "After the shock and chaos of the terrorist attack on the U.S. on Tuesday, a vast emergency response plan has been activated involving thousands of federal and local authority staff, and medical, social care and public health workers. Within hours of the catastrophe, president George Bush formally declared it a major disaster, triggering the release of a massive federal disaster programme of resources and funding for New York and Washington. Scores of government and local authority departments, voluntary sector organisations and health agencies are involved, coordinated by the Federal Emergency Management Agency (FEMA) in Washington."

Journalists emphasized the magnitude of the attack, the difficulty of the rescue operation, and the fact that the attacks were the "crisis of Bush's presidency." The September 11 attacks were also repeatedly compared to Pearl Harbor, and it was suggested that retaliation was likely to follow in the coming weeks:

> BBC anchor, interviewing correspondent: "Now, one prominent republican senator from Washington, Senator Chuck Hagel, is saying this is the second Pearl Harbor. I don't think that I'm overstating this. I mean, is that the sense that you have on the streets of New York, that this feels like a city which has suffered an act of war?"

British journalists also frequently speculated that more attacks were coming, even though they had no evidence to support that claim. Some even suggested that terrorists were planning to use nuclear, chemical, and biological weapons:

> London Times Sept. 13: "There is, alas, worse to come, even worse than the destruction of the World Trade Centre and the attack on the Pentagon. The over-concentration in recent intelligence assessments on the coming threat from weapons of mass destruction does not mean that this threat is illusory. The threat is real. The question, alas, is not whether the terrorists of the 21st century will use weapons of mass destruction, but when and where they will do so."

Just as with the British coverage of September 11, a variety of reports in the United States were given about how calmly people dealt with the crisis. Although some statements reflected a relative degree of calm, others called the situation "chaotic" and "panicked":

> CNN anchor: "This is one of those situations that is extraordinary chaotic. Even in the best of planning, I think it's fair to say that it is chaotic. And officials are trying to do many things at one time."

Much of the breaking TV news coverage in the United States assured viewers that emergency workers and response teams were doing the best that they could and reassured audiences that the government was continuing to carry on:

CNN reporter: "But as a city, you know, we come together, and our emergency services provide every support they can in the face of such a senseless tragedy."

There was also a sense from the U.S. coverage that New York City and
Washington, D.C., would have great difficulty returning to normal after the
attacks, suggestions that America would never be the same again and references to this being the worst day in American history:

USA Today Sept. 12: "Days that live in infamy are supposed to be found in dusty history books. Tuesday changed all that. It changed everything. Our world will never be
the same."

Let's now turn to the second research question: the general tenor of the
British coverage of 7/7 was that London would quickly return to normal,
unlike the U.S. coverage of September 11, which suggested the opposite. By
running its normal weather report soon after the story broke, the BBC offered
further proof that one of its objectives was to calm the public and project a
sense of normalcy. After the initial report of the explosions, the BBC returned
to regular programming for more than twenty minutes before beginning continual live coverage. The BBC continued to take breaks to report the weather
throughout the crisis coverage. However, these breaks for regular programming indicate to some in the United Kingdom that the BBC is not the place to
turn during a crisis when you want information first because commercial
broadcasters were more forthcoming with the details of the terrorist attack.
The BBC also faced claims of political correctness related to the Muslim terrorist suspects who were arrested in summer of 2006. An independent report
commissioned by the BBC to analyze the network's performance quoted one
member of the public surveyed for the report who claimed, "I think the BBC
is too politically correct. The BBC were saying '21 men have been arrested'
and I thought 'what's happening?' So I flicked over to Sky and it says '21
Asian men have been arrested.'"

Even during the breaking news reports, the incident, although sometimes
described as chaotic, was also portrayed as controllable and within the capacity of the emergency services that had been preparing for such an attack for
years:

BBC reporter: "Well I think the most notable thing is a complete absence of panic and
a great deal of stoicism. No one shouting or particularly panicking."

The Guardian July 8: "Tony Blair last night praised the 'stoicism and resilience' of Londoners in the face of yesterday's onslaught on the capital's transport system by bombers he implied were Islamist terrorists."

The majority of the British 7/7 coverage was devoted to sending a message that daily life was returning to normal fairly quickly:

The Guardian July 9: "London was almost back to normal yesterday. Most of the public transport system was functioning and Sir Ian urged everyone to be behind their desks at the start of next week. "This is business as usual on Monday," he said. "We go on.""

The American coverage of the July 7 bombings drew parallels with 9/11, but highlighted key differences, including the relative calmness with which the London event was handled. American journalists emphasized that London was prepared for the attack, and there was little panic, although sometimes it seemed as if the journalists expected more panic than there was:

CNN reporter: "And of course, this is London. This is a city more than perhaps any other modern European city that has dealt with terrorist attacks over the last several decades, throughout the height of the troubles with the IRA, the Irish Republican Army. The British paramilitary services, the police, the army, and all the emergency services are extremely well trained and familiar with this kind of attack, and they have sprung into action."

New York Times July 11: "For several heart-stopping hours, it seemed as if last Thursday would be Britain's 9/11. There were the same panicky television reports—an explosion here, a power outage there, a bomb somewhere else—and the same terrible sense that events were unfolding almost too quickly for human understanding, that there might be fresh hell yet to come."

CNN reporter: "This is going to create enormous panic and terror within London because, obviously, it is going to disrupt the economy, it's going to disrupt the office life, ad people are going to be very frightened."

U.S. newspaper coverage emphasized how quickly life returned to normal and discussed the resilient response of Londoners to the attacks:

USA Today July 8: "Over the space of 56 minutes, at the height of rush hour, this scene of death and destruction from an apparent terrorist attack would be repeated on two more subway trains and a red double-decker bus, leaving at least 37 people dead and 700 wounded. It was the worst attack on London since World War II. By early evening the city had taken on a calmness, though Londoners faced a tense end-of-the-workweek today. A drizzle stopped, and people walked home from work or sipped a pint of beer in a pub with friends. Bus service in the city center was restored. Police tape still protected bombed areas, but life around much of the city seemed normal: A group of young men played soccer in a park near Russell Square."

The tone of the coverage also varied by country reporting. Common adjectives during the U.S. coverage of the British subway bombings included "fear," "terror," and "helpless." Common adjectives from the British press were "calm," "resilient," and "defiant." The British press emphasized the unusually low number of deaths, considering the type of attack, and talked about massive criminal investigation. The U.S. press emphasized war language, much in the same way it did in its September 11 coverage. The reporting then quickly turned to "Could it happen in the United States" type stories and stories that focused on the search for any Americans hurt in the bombings. Statements about future attacks during the British coverage of 7/7 were rare. Only a few newspaper articles speculated that there were more attacks likely to come, or that Britons had anything to fear from new attacks:

> The Guardian July 9: "The security and intelligence agencies are also expressing concern that extremists trained in Iraq will come to–or return to–Britain with new bomb-making skills."

Research question three compared the coverage of 9/11 and 7/7. One of the key differences between the American and British coverage of September 11 and the July 7 subway bombings is in how often a critical view was employed in the coverage. The British news media were far more critical of every aspect of the 9/11 and 7/7 attacks, from the response of the government to the political implications of potential retaliation to the analysis of U.S. foreign policy decisions that could have led to the attacks. The British media provided a deeper perspective that put the attacks into a political context. The U.S. government was criticized for being unprepared for the attacks, for ignoring intelligence warnings, for not responding fast enough, and for not having adequate security precautions in place:

> The Guardian Sept. 13: "The CIA, the FBI, and America's national security agency spend billions of pounds a year gathering intelligence abroad and combating terrorism at home–significantly more than any other country. Their satellites can spot vehicle number plates and eavesdrop on millions of faxes, emails, and telephone calls. Yet

they failed to prevent audacious attacks on hugely symbolic American landmarks by terrorists who must have spent a long time in the U.S. preparing them."

Other reports cited the lack of understanding in the United States about issues in the Middle East:

The Guardian Sept. 12: "Francois Loncle, chairman of the foreign affairs committee of the French parliament, argued that the attacks were the result of an underestimation by the west of the problems in the Middle East. Washington, he said, should pay 'a little more and a little better attention to world affairs.'"

Despite the fact that it happened on their soil, the British coverage of 7/7 was quite critical. Journalists questioned whether the entire Tube network should have been closed down, criticized the emergency response teams for not responding quickly enough, and questioned the British government for its involvement with the war in Iraq that could have prompted the attacks. Journalists also discussed how British policies might have contributed to the attacks while also suggesting that the attacks in London were not preventable.

BBC reporter: "So both politicians and security people have been saying for a long time Britain is a target and that's inevitable."

The U.S. media also took a more critical view in the coverage of 7/7 than they did during the 9/11 coverage. The most common critical angle was how this attack could have happened despite London's extensive network of security cameras. The American coverage also criticized the American government for not doing enough to prevent this type of attack from occurring on American transportation networks. It was also emphasized in the U.S. press that Britain had been expecting this type of attack for some time:

New York Times July 8: "Police and intelligence officials acknowledged that they were taken completely by surprise by the coordinated bombings, even though they had been anticipating a terrorist attack for years."

CNN anchor: "So there's a fair amount of security, and many people pointed out that already the city was already at high alert. SO within all of that, to some degree, it points out that maybe it is impossible to really prevent any kind of attack like this."

American coverage also frequently questioned why the threat level in the United States was not raised as a result of the incidents in London:

CNN anchor: "Here in the U.S., though, the threat level has not been raised out of Washington, D.C, although officials are quick to note that there are ongoing discussions about raising it. That seems a little bit surprising to me."

However, CNN rarely took a critical view of the events of September 11 in the immediate aftermath of the attacks. The few critical statements came from officials who were interviewed who criticized the government's lack of intelligence. The U.S. in-depth newspaper coverage that followed the attacks was much more critical, not only of the intelligence failure that led to the attacks and speculation that the government had advanced warnings, but also of the potential response of the United States and what effect that would have on American foreign policy:

USA Today Sept. 12: "Intelligence officials said that for at least the past 10 days they had anticipated a possible attack by followers of Osama bin Laden, the Saudi-born financier of Islamic terror groups. But the officials said they had expected the attacks, if any, to come against American targets abroad. American embassies and military installations overseas were on high alert Tuesday."

Other U.S. newspaper articles were quick to criticize the way President Bush handled the attacks, claiming that by not immediately returning to Washington he gave the impression of being afraid. The newspapers also ran articles about Europe's own critical view of a military response:

USA Today Sept. 18: "But since then, some NATO members have expressed reservations about military strikes that could provoke an anti-Western backlash in the Muslim world. French Defense Minister Alain Richard warned Monday that military action might destabilize an already shaky region."

U.S. reports of 9/11 often speculated that terrorists have or would soon have access to biological, chemical, or nuclear weapons, and the attacks themselves were even sometimes described as if these types of weapons were used. The New York Times reported on September 12: "Many experts said they foresaw a growing likelihood of large-scale terrorist events. But they said that was based on the fear that well-organized and well-financed groups were increasingly close to obtaining weapons of mass destruction, including biological and nuclear weapons." During the breaking news coverage a CNN reporter, watching the tower collapsing said, "And I can't—I tell you that I can't see that second tower. But there was a cascade of sparks, and fire, and now this— it looks almost like a mushroom cloud explosion."

Both the British and U.S. coverage of September 11 focused on retaliation rather than criminal investigation, although British statements were not as powerful as they were in the American media. But, both the U.S. and British media framed the attacks as acts of war and suggested that immediate retaliatory action should be taken. On September 16, an article in the *London Times* stated, "as the towers fall a manhunt begins. Out of the dust, a war declaration."

The war sentiment was emphasized much more in the U.S. coverage where CNN, *The New York Times*, and *USA Today* all framed the 9/11 attacks in a way that made retaliatory military action both justified and necessary. The U.S. coverage reflected a message that this was an act of war that demanded military action as a response:

> *New York Times* Sept. 12: "A growing number of officials said the magnitude of today's attacks put them beyond the reach of law enforcement. They said that arrest and trial of conspirators was an inadequate response to what amounted to an assault on the nation's security that could be dealt with only by military force ... The re-evaluation of how the United States should respond was apparent in Mr. Graham's remarks. He said that in response to the attacks he would be willing to reassess the government's ban on assassinations of foreign leaders."

The *New York Times*, in particular, reported many of the brash statements from lawmakers that suggested a full-blown military response, as well as statements from ordinary Americans reflecting intense anger and a desire to retaliate:

> *New York Times* Sept. 14: "Phil Beckwith, a retired truck driver, announced his modest proposal for avenging the attacks on New York and Washington in the editorial offices of The Ranger, a newspaper that serves Fremont County, Wyo., one of the largest and emptiest counties in the nation. He had gone to the paper to buy a classified advertisement."

> "'I know just what to do with these Arab people,' Mr. Beckwith proclaimed on Wednesday to the newspaper staff. 'We have to find them, kill them, wrap them in a pigskin and bury them. That way they will never go to heaven. Now, I would like to buy an ad to rent my house.'"

The British coverage of 7/7 was far more straightforward and clearly focused on a criminal investigation. There were no suggestions of retaliatory action, but there was information about how the government and the prime

minister were dealing with the crisis, as well as what security measures were
in place to prevent the likelihood of another attack:

> BBC reporter: "It is likely to be very intense activity in Whitehall. COBRA, which is
> the emergency committee that has been called, they've been meeting to determine
> standard procedure in any major incident, terrorist or not. You should also expect the
> Joint Terrorist Assessment Center, J-TAC, which is housed in MI5, will be looking at
> what information is coming in, maybe recently, whether there is evidence of who has
> carried this out."

A similar tone was found in the U.S. coverage of 7/7, which highlighted
how the British government was handling the crisis and how well-prepared
Britain was to conduct the investigation given its past experience with terrorism:

> CNN anchor: "Tony Blair, the prime minister of great Britain, addressed reporters at
> Gleneagles, Scotland, where that G8 Summit is under way, indicating his desire, his
> intent to get back to London as quickly as possible, spend the day there, get a full re-
> port, and then return back to the G8, so the summit can continue on."

> New York Times July 8: "Britain has considerable experience investigating bombs and
> identifying those responsible, based on years of attacks in London and in Northern
> Ireland by the Irish Republican Army."

During 9/11 coverage, both the British and U.S. coverage employed nationalistic statements, but to a larger degree in the U.S. press. The U.S. coverage focused on the symbolism of the attacks, the ability of the American
people to pull together and recover, and expressions of patriotism and support
for the government and its leaders. There was a great emphasis on politicians
putting aside their political differences and uniting to fight terrorism. There
were also repeated mentions of "American resolve" and determination. The
American media coverage of September 11 was filled with patriotism and
strong support for the government and emphasized the international support
as the New York Times noted in an article published on September 16: "Of all the
offers of support that Americans received last week from friends and acquaint-
ances abroad, one of the most remarkable came from the editors of Le Monde,
the French daily newspaper often noted for wariness of the United States. 'We
are all Americans,' the paper declared in a banner headline."

Nationalistic statements in the British coverage of 9/11 focused on what
the British government was doing in response, and how many Britons had
been killed:

The Guardian Sept. 13: "More details are emerging of Britons killed, safe and missing after the terrorist attack on New York, following home secretary Jack Straw's confirmation of almost 100 British deaths."

BBC anchor: "As a sister city to New York, London is sharing some of the shock being felt across the Atlantic. And as it sinks in we are increasingly turning our thoughts to how we would cope if it happened here."

During their 9/11 breaking news coverage, the BBC cut to local news coverage, program commercials, and the weather several times. At first glance this might perhaps suggest an enthnocentric view because the attack was not domestic, but as noted earlier, the BBC reverted to regular reporting during the 7/7 attacks as well. However, much of the 7/7 coverage in the United States did seem ethnocentric, highlighting the American response to the tragedy—what should be done in the United States to prevent the attack from happening here, how many Americans were killed, how President Bush responded to the attack. Former New York City mayor Rudolph Giuliani was apparently down the street from the attacks as they were happening. This was a point given tremendous attention in the American coverage of the July 7 attacks:

New York Times July 8: "Rudolph W. Giuliani, whose legacy as mayor of New York City was transformed by his stewardship during and after the Sept. 11 terrorist attacks, was on a business trip to London yesterday when he found himself half a block from the first blast there."

Many newspaper articles in the United States much like the coverage in London, centered on the stoicism of the British people as they dealt with the crisis:

New York Times July 11: "He pointed to the crowds, thousands of people gathered around Buckingham Palace and down the large boulevard in front it, defying the notion that such a large gathering could prove a tempting terrorist target. 'There's something about Londoners–about the British,' he said. 'We just tend to get on with it.'"

The Guardian July 12: "In another letter addressed directly to the terrorists, an anonymous Londoner had written: 'If you are looking to boost morale, our pride, then you have succeeded. If you want to ensure our commitment to our way of life you have achieved much. If you expect people to crawl out of smoke-filled tunnels, head to work and otherwise get on with their daily lives, you were right. If your aim was to raise our strength and defiance, congratulations.' The letter ended with the rhetorical question and retort: 'Burning with fear? Not bloody likely.'"

Just as the BBC cut to program commercials and weather updates during their 9/11 coverage, CNN also covered domestic news during its July 7 coverage, including a report on missing teenager Natalee Holloway, and a weather update on Hurricane Dennis.

The British and American coverage of 7/7 and 9/11 contained important differences. These differences stem from separate British and American ideas about the media's role in society, different press models, cultural and historical values, and proximity to the event reported. The British media's traditional role of reporting for the good of the public was reflected in the coverage. Britain's past experience with terrorism and cultural tradition of stoicism contributed to a difference in the language and tone between the American and British reports of the attacks. This difference in tone was best illustrated in the ways that the press calmed the public and/or caused fear. The American media coverage caused fear more often than the British coverage. On the other hand, the British coverage spent more airtime and column inches calming the public. For example, although neither country made any attempt to underscore the magnitude of the 9/11 attacks, the British coverage did a much better job of reassuring the public that though the attacks were horrific, life would go on. This attitude reflects a typically British response to terrorism that was often employed during past terrorist incidents with the IRA. The British journalists reporting on September 11 were quick to draw parallels with their own past terrorist incidents, and frequently made reference to the fact that the United States had so little experience with domestic terrorism. The overall calmness of the British coverage of the July 7 bombings was also noticeable. Throughout the coverage there was a sense that the incident was under control, the emergency services were well prepared to deal with the crisis, and life would return to normal as quickly as possible. Reassuring the public appeared to be a top news priority. Reporters kept referencing the emergency plan that had been in place in London, and how the authorities had rehearsed for such an occasion.

Also interesting to note was a British fatalistic attitude toward the attacks. They were inevitable; there was nothing that could have been done to prevent them, and now was the time to put on a brave face and deal with it. The BBC coverage was so focused on carrying on with life as usual that they broke away from continuous coverage to show a weather report. Several factors likely contributed to this calmness, including the British cultural attitude of stoicism, which was a repeating theme in the coverage. Reporters frequently made reference to the British "stiff upper lip" and the resilience of the people of London. This was best illustrated in a clip, broadcast several times, of an elderly man on the street being interviewed by a reporter. "Well, we've been here before, haven't we?" he asked. The British people have had their fair share of violence on their home turf, from air raids during World War II to the troubles with the IRA. The idea that there are people out there who wanted to attack them did not produce the same shocked and humiliated response as it did on the other side of the Atlantic, simply because Britons had "been there before."

The key difference in the U.S. 9/11 coverage was the emphasis placed on fear and the end of normalcy. The American media framed the attacks of September 11 as an act of war that made military retaliation not only justified, but also necessary. Both television and newspaper coverage spread the message that life in America would never be the same and that fighting terrorism was the new war of the twenty-first century. Bold statements from leaders and lawmakers about mercilessly hunting down and killing terrorists filled pages of newspapers in the weeks following the attacks. Perhaps, this coverage targeted a public already on board with this notion. Polls printed in the newspapers reflected the overwhelming support of a retaliatory strike. The U.S. media repeated analogies to nuclear weapons and the "apocalypse." Ground Zero was described as a scene from a horror movie where Lower Manhattan was recovering from nuclear winter. Both television and print journalists suggested that terrorists had access to nuclear, chemical, and biological weapons, and it wasn't a matter of if they would use them, but *when*. The American coverage of September 11 did little, if anything, to calm the public. Instead, it encouraged viewers and readers to prepare themselves for a war.

Coverage of the British government response to the July 7 bombings differed from the American government response to September 11. Analysis of the British government's response to the bombings in both the American and British media contained no mention of retaliation or military deployment. This contrasted sharply with the American media's war speculation in the weeks following September 11. When discussing the investigation and response to the July 7 attacks, the British media treated the inquiry into the attacks as a criminal investigation rather than a military investigation. The BBC, in particular, was very hesitant to even use the word "terrorism" in their reports, calling the terrorists "attackers" or "bombers." As noted in chapter 2, the BBC has specific editorial guidelines for reporting on terrorism and maintains that acts of terror must be reported responsibly, avoiding words that carry "emotion or value judgments."

The American coverage of September 11 was also different from the British coverage in the area of critical analysis of the government. Both CNN and the two American newspapers used in this study were far less likely than their British counterparts to criticize the response and actions of the government and its leaders, or to provide analysis of foreign policy and terrorist motives. The language used by British reporters was reminiscent of earlier BBC news reports during the troubles with the IRA. Terrorist motives were discussed and analyzed in far greater detail than in the American coverage. While British journalists emphasized that the attacks were designed to disrupt the daily lives of Americans and spread a message of discontent with American policies in the Middle East, the American reporters tended to oversimplify or dismiss the intentions of terrorists, or simply not discuss them at all. Relatively little political context was given to explain to readers and viewers what actions or policies of the United States could have contributed to the attacks. While the BBC and British newspapers were expressing the need for caution and diplomacy in response to the attacks, CNN, the New York Times, and USA Today were creating an image of a battle between good and evil, a conflict that would undoubtedly lead to a war.

The British media, in particular the newspapers, were also much more critical of leaders and government than the American media after September 11. The Guardian and the London Times were highly critical of British foreign policy, including the war in Iraq and the strong relationship between former prime minister Tony Blair and President Bush. The British press had a much greater tendency to question the actions of leaders rather than unconditionally support them, as was more common in the American press after September 11.

These two different interpretations of the same event have several possible explanations. One reason for the American media's lack of restraint, criticism, and analysis during the September 11 attacks in comparison to the British coverage was that the September 11 attacks remain the deadliest on U.S. soil. Further, Americans had relatively little experience in dealing with domestic terrorism. The attacks took the country by surprise, not only because intelligence failed to detect them, but also because Americans had been living under the assumption that they were immune to attack. Given this combination of factors, it is not surprising that the American response was one of shock, anger, a desire for retaliation, and an overwhelming support for the government and its leaders.

The most striking contrast in coverage comes from comparing television coverage from the United States on 9/11 to British television coverage on 7/7. Here the functional differences between the BBC and CNN were the most clear. The BBC emphasized advice to Londoners about how to get home, where to pick up their children from school, and most importantly, orders to remain calm were repeated throughout the broadcast. The calming and informative tone contrasted sharply with CNN's more confused and chaotic account of September 11. This is even more striking considering that these are journalists reporting on a crisis in their own country. This is not to say that CNN neglected its duties in reporting September 11. It is important to emphasize that the nature and scope of these two attacks was dramatically different.

The different press models may also have influenced coverage. The BBC placed a greater emphasis on social responsibility and public service. John Reith, who served as the first director general of the BBC, and is credited with the establishment of the tradition of public service broadcasting in Britain, rejected the idea that it is the responsibility of the broadcaster to give the public what they want. He said, "Public service is the maintenance of high standards, the provision of the best and the rejection of the hurtful ... [I]t is occasionally indicated to us that we are apparently setting out to give the public what we think they need—and not what they want—but few know what they want and very few what they need—in any case it is better to overestimate the mentality of the public than to underestimate it."[36] In the United States, broadcasters were not quick to agree with Reith's public service model. In 1933, the National Association of Broadcasters produced a pamphlet that asserted, "The nervously active American is never in a mood to take educational punishment. You must interest him—or he quickly tunes you out. This characteristic is in only slightly lesser degree fundamental to any discussion of listener reaction in any country. It is the rule and the law and the testament upon which every successful broadcast structure is based. It is the risk, for instance, that Sir John Reith runs in Britain when he avowedly gives his public what he believes it is good for it to have."[37] The idea that the media should give out information that is in the best interest of the viewer has cultural and historical roots in Great Britain. They have "spent years living under various permutations of socialist government. This has created different habits of mind, and softened the collective rhetoric ... Brits of most persuasions are

happiest talking about 'self-reliance' and 'the common good,' which remind
them of the War, the Crown, and the BBC in no particular order."[38]

Conclusion

This case study illustrates some of the ways that the British and U.S. press
cover terrorism differently, and these differences come from a variety of cultural factors. The more calm and measured British media reaction to both attacks not only reflected the British tradition of reserved stoicism, but also
contrasted sharply with the fear-causing American coverage, which generally
made a far less solid attempt to disseminate crucial information to viewers and
readers. The British coverage, particularly the BBC coverage of the July 7 London Underground bombings, was notable for its conservative news judgment
that prevented panic and assured viewers that the situation was under control.
The British coverage, both newspaper and television, also did a good job of
providing important political context such as past foreign policy decisions in
the Middle East, which were largely absent in the American coverage. In-
depth consideration of possible root causes of terrorism reflects both the British media's experience with terrorism and, in the case of the BBC, adherence
to documented standards for reporting terrorism.

The American coverage's lack of political context was alarming when considering what it was replaced with: oversimplification of terrorist goals, intense elements of patriotism and national pride, and perhaps most
significantly, unwavering support for government. The American media, in a
sense, abandoned their role as a watchdog in the wake of the September 11
attacks.

Notes

1   This chapter is based on data from research conducted during the following two studies
    presented as conference papers: Tayler Kent and Brooke Barnett, "Framing Terrorism: Brit-
    ish and U.S. Coverage of 7/7 and 9/11" (Paper presented to the news division of the an-
    nual meeting of the Broadcast Education Association in Las Vegas Nevada, April 2007);
    Brooke Barnett, Amy Reynolds, Laura Roselle, and Sarah Oates, "Journalism & Terrorism
    across the Atlantic: A Qualitative Content Analysis of CNN and BBC Coverage of 9/11 and
    7/7" (Paper presented at the Association for Education in Journalism and Mass Communi-
    cation annual meeting, Washington, D.C., August 2007).
2   Peter Huck, "We Had 50 Images within an Hour,'" Guardian, July 11, 2005.
    http://www.guardian.co.uk/attackonlondon/story/0,16132,1525911,00.html.

3   Robert G. Picard, *Media Portrayals of Terrorism: Functions and Meaning of News Coverage* (Iowa: Iowa State University Press, 1993); Brigitte L. Nacos, *Mass-Mediated Terrorism: The Central Role of the Media in Terrorism and Counterterrorism* (Lanham, Md.: Rowman & Littlefield, 2007).

4   Robert G. Picard and Paul D. Adams, "Characterization of Acts and Perpetrators of Political Violence in Three Elite U.S. Daily Newspapers," *Political Communication and Persuasion*, 4 (1987): 1–9.

5   Matthew A. Baum, *Soft News Goes to War: Public Opinion and American Foreign Policy in the New Media Age* (Princeton, N.J.: Princeton University Press, 2005).

6   Philip Seib, "The News Media and the 'Clash of Civilizations,'" *Parameters*, 34 (2004): 71–85.

7   Nacos, 223.

8   Kristen Alley Swain, "Outrage Factors and Explanations in News Coverage of the Anthrax Attacks," *Journalism and Mass Communication Quarterly*, 84, 2 (2007): 347.

9   Philip Schlesinger, "Graham Murdock and Philip Elliott," *Televising Terrorism: Political Violence In Popular Culture* (London: Comedia, 1983).

10 Robert Samuelson, "Unwitting accomplices," *Washington Post*, November 7 2001, A29.

11 Raphael Cohen-Almagor, "Media Coverage of Acts of Terrorism: Troubling Episodes and Suggested Guidelines," *Canadian Journal of Communication*, 30 (2005): 3.

12 David L. Altheide, *Terrorism and the Politics of Fear* (Lanham, Md.: AltaMira Press, 2006); Robert M. Entman, "Framing U.S. Coverage of International News: Contrasts in Narratives of the KAL and Iran Air incidents," *Journal of Communication*, 4, 4 (1991): 6–27.

13 Sarah Oates, "Comparative Aspects of Terrorism Coverage: Television and Voters in the 2004 U.S. and 2005 British Elections" (Paper presented at the Political Communication Section pre-APSA conference, Annenberg School for Communication, University of Pennsylvania, August 2006).

14 National Public Radio, "Roundtable: Egypt Terrorism Attacks, AFL-CIO Troubles," aired July 25, 2005, http://www.npr.org/templates/story/story.php?storyId=4769688.

15 Einar Östgaard, "Factors Influencing the Flow of News," *Journal of Peace Research*, 2, 1 (1965): 39–63.

16 Stephanie Craft and Wayne Wanta, "U.S. Public Concerns in the Aftermath of 9/11: A Test of Second Level Agenda Setting," *International Journal of Public Opinion Research*, 16, 1 (2004): 456–463.

17 Catherine A. Luther and Xiang Zhou, "Within the Boundaries of Politics: News Framing of SARS in China and the United States," *Journalism & Mass Communication Quarterly*, 82, 4 (2005): 857–872.

18 Philip Schlesinger, Graham Murdock, and Philip Elliott, *Televising Terrorism: Political Violence in Popular Culture* (London: Comedia/Marion Boyars. 1983).

19 Entman.

20 Elisabeth Anker, "Villains, Victims and Heroes: Melodrama, Media, and September 11," *Journal of Communication*, 55, 1 (2005): 22–37; Sandra L. Borden, "Communitarian Journalism and Flag Displays after September 11: An Ethical Critique," *Journal of Communication Inquiry*, 29, 1 (2005): 30–46; Amy Reynolds and Brooke Barnett, "'America under Attack' CNN's Visual and Verbal Framing of September 11," in Steven Chermak, Frank Bailey, and Michelle Brown, eds. *Media Representations of September 11th* (New York: Praeger, 2003), 85–101.

21 Anker, 22–37.

22 Reynolds and Barnett, 85–101.

23 Altheide.

24 Michael Ryan, "Framing the War against Terrorism," *International Communication Gazette*, 66, 5 (2004): 363–382.

25 Dahr Jamail, E-Mail Message to the Author, March 10, 2008.

26 Daniela Dimitrova, Lynda Lee Kaid, Andrew Paul Williams, and Kaye D. Trammell, "War on the Web: The Immediate News Framing of Gulf War II," *Harvard International Journal of Press/Politics*, 10, 1 (2005): 22–44.

27 Sean Aday, Steven Livingston, and Maeve Hebert, "Embedding the Truth: A Cross-cultural Analysis of Objectivity and Television Coverage of the Iraq War," *Harvard International Journal of Press/Politics*, 10, 1 (2005): 3–21.

28 Entman, 52.

29 W. Russell Neuman, Marion R. Just, and Ann N. Crigler, *Common Knowledge: News and the Construction of Political Meaning* (Chicago: University of Chicago Press, 1992).

30 Hsiang Iris Chyi and Maxwell McCombs, "Media Salience and the Process of Framing: Coverage of the Columbine School Shootings," *Journalism and Mass Communication Quarterly*, 81, 1 (2004): 22–35.

31 Paul Adams, "American Interest in Iraq Slumps," BBC News Monday, March 24, 2008, http://news.bbc.co.uk/2/hi/middle_east/7311814.stm.

32 Danny Schechter, *Media Wars: News at a Time of Terror* (Lanham Md.: Rowman & Littlefield, 2003).

33 Analysis conducted from data from http://people-press.org/nii/.

34 Reynolds and Barnett, 689–703.

35 Holli A. Semetko and Patti M. Valkenburg, "Framing European Politics: A Content Analysis of Press and Television News," *Journal of Communication*, 50 (2000): 93–109.

36 William L. Rivers, *Responsibility in Mass Communication* (New York: Harper and Row, 1980).

37 Michele Hilmes, "British Quality, American Chaos," *Radio Journal: International Studies in Broadcast and Audio Media*, 1, 1 (2003): 13–27.

38 Jane Walmsley, *Brit-Think Ameri-Think* (New York: Penguin, 1986).

CHAPTER 7

The Challenge of Patriotism
WHEN JOURNALISM IS ACCUSED OF TERRORISM

To announce that there must be no criticism of the President or that we are to stand by the President, right or wrong, is not only unpatriotic and servile, it is morally treasonable to the American public.
—President Theodore Roosevelt

It was almost the end of the program. CNN's Wolf Blitzer was talking about the Iraq War with Richard Perle, who served as the assistant U.S. defense secretary during the Reagan administration and on the advisory Defense Policy Board during both Bush administrations. The discussion lead to Seymour Hersh, the journalist best known for his investigation of the My Lai massacre in Vietnam. In 2004, Hersh broke the major story about mistreatment of prisoners at Abu Ghraib prison in Iraq. Blitzer held the newest copy of the *New Yorker* and said, "There's an article in the *New Yorker* magazine by Seymour Hersh that's just coming out today in which he makes a serious accusation against you that you have a conflict of interest in this because you're involved in some business that deals with homeland security, you potentially could make some money if, in fact, there is this kind of climate that he accuses you of proposing. Let me read a quote from the *New Yorker* article, the March 17th issue, just out now. 'There is no question that Perle believes that removing Saddam from power is the right thing to do. At the same time, he has set up a company that may gain from a war.'"

Richard Perle responded, "I don't believe that a company would gain from a war. On the contrary, I believe that the successful removal of Saddam Hussein, and I've said this over and over again, will diminish the threat of terrorism. And what he's talking about is investments in homeland defense, which I think are vital and are necessary. Look, Sy Hersh is the closest thing American journalism has to a terrorist, frankly."

Blitzer started to ask a new question and then processed what Perle had just said, "Well, on the basis of—why do you say that? A terrorist?" Perle responded, "Because he's widely irresponsible. If you read the article, it's first of all, impossible to find any consistent theme in it. But the suggestion that my views are somehow related for the potential for investments in homeland defense is complete nonsense."

Blitzer, who seemed utterly confused, said, "But I don't understand. Why do you accuse him of being a terrorist?" Perle again attempted to explain,

"Because he sets out to do damage, and he will do it by whatever innuendo, whatever distortion he can—look, he hasn't written a serious piece since My Lai."

It was then time to close the show and Blitzer wrapped up: "All right. We're going to leave it right there. Richard Perle, thank you very much." And so a well-known and celebrated journalist was called a terrorist by a high ranking government official simply because the official disagreed with him.

Although it is rare for a former U.S. government employee to call an American journalist a terrorist, more subtle names and insinuations often arise, such as suggesting that a journalist is behaving in "un-American" ways simply for reporting critically on the government. Journalists often talk about how these insinuations affect what gets covered. On a 2003 Topic A with Tina Brown program on CNBC, CNN reporter Christiane Amanpour said she felt that government claims that the press behaved unpatriotically in airing stories that questioned some decisions made by the Bush administration did ultimately impact coverage of the Iraq war that year. "I think the press was muzzled, and I think the press self-muzzled," she said. "I'm sorry to say, but certainly television and, perhaps, to a certain extent, my station was intimidated by the administration and its foot soldiers at Fox News. And it did, in fact, put a climate of fear and self-censorship, in my view, in terms of the kind of broadcast work we did."

Host Tina Brown then asked Amanpour if there was any story during the war that she couldn't report. "It's not a question of couldn't do it; it's a question of tone," Amanpour said. "It's a question of being rigorous. It's really a question of really asking the questions. All of the entire body politic in my view, whether it's the administration, the intelligence, the journalists, whoever, did not ask enough questions, for instance, about weapons of mass destruction. I mean, it looks like this was disinformation at the highest levels." In response, Fox News spokeswoman Irena Briganti said of Amanpour's comments: "Given the choice, it's better to be viewed as a foot soldier for Bush than a spokeswoman for al-Qaeda."

These exchanges epitomize the challenge of journalism during crisis. Well-respected journalists are framed as aiding and abetting terrorism when they critically examine government officials and policies. Hersh's work has won more than a dozen major journalism prizes, including the Pulitzer Prize for International Reporting and four George Polk Awards. Amanpour has won an Emmy, two George Foster Peabody Awards, two George Polk Awards, a

Courage in Journalism Award, a Worldfest-Houston International Film Festival Gold Award, and the Livingston Award for Young Journalists. Hersh and Amanpour are impressive and important investigative and international journalists. And, as the above exchanges show, in some instances their journalistic work was likened to terrorism or support for terrorists. Such imprecise usage of the term further confuses what terrorism actually is. As Dr. Ariel Merari, head of the Center for Political Violence at Tel Aviv University, has said, when the term terrorism becomes synonymous for generally negative behavior, its usefulness is only "in propaganda."[1] In this case, the term was used propagandistically to establish a simplistic view of fighting terrorism—if you disagree with the government's response to terrorism, then you are with the terrorists. Journalists who question the government are seen as aligning themselves with the terrorists. A similar fate has been found for words such as fascist or heinous.

Patriotism

Pew Research studies have always found strong support for the notion of patriotism. The 2007 survey, in line with the surveys for the past twenty years, shows that 90 percent of the public concur with the statement, "I am very patriotic."[2] A University of Chicago study found that Americans are the most patriotic citizens in the world.[3]

However, what is meant by individuals who say they are patriotic is often harder to assess. Scholar Robert Jensen writes:

> If we use the common definition of patriotism—love of, and loyalty to one's country—the first question that arises is, what is meant by country? Nation-states, after all, are not naturally occurring objects. What is the object of our affection and loyalty? In discussions with various community groups and classes since 9/11, I have asked people to explain which aspects of a nation-state—specifically in the context of patriotism in the United States—they believe should spark patriot feelings ... The answers offered include the land, the people of a nation, its culture, the leadership, national policies, the nation's institutions, and the democratic ideals of the nation."[4]

Leo Tolstoy wrote and spoke often about patriotism and noted that "It is generally said that the real, good patriotism consists in desiring for one's own people or State such real benefits as do not infringe the well-being of other nations."[5]

Patriotism is often criticized.[6] Critics of patriotism find that preferring one's own country over others is morally problematic in that patriotism suggests the superiority of a whole nation.[7] Anarchists and socialists often wrote about how patriotism was a tool of oppression used against the military and working class. Emma Goldman, writing in the early 1900s, said:

> Thinking men and women the world over are beginning to realize that patriotism is too narrow and limited a conception to meet the necessities of our time. The centralization of power has brought into being an international feeling of solidarity among the oppressed nations of the world; a solidarity which represents a greater harmony of interests between the workingman of America and his brothers abroad than between the American miner and his exploiting compatriot; a solidarity which fears not foreign invasion, because it is bringing all the workers to the point when they will say to their masters, "Go and do your own killing. We have done it long enough for you."[8]

Scholars have likened patriotism to racism in that it requires a blind preference for one group over another.[9] Others have contended that patriotism is fundamentally at odds with liberal morality.[10] Samuel Johnson provides perhaps the most famous critical comment of patriotism in that it is the "last refuge of a scoundrel."[11] But, he also spoke specifically about the press in relation to patriotism:

> In a time of war the nation is always of one mind, eager to hear something good of themselves and ill of the enemy. At this time the task of the news-writer is easy; they have nothing to do but to tell that a battle is expected, and afterwards that a battle has been fought, in which we and our friends, whether conquering or conquered, did all, and our enemies did nothing.[12]

The argument against a patriotic press is that the press needs to be able to stand outside of an allegiance to one's country and report on key issues, especially during times of war. Robert Jensen wrote about patriotism and the press after the September 11 attacks:

> Two definitions competed after the terrorist attacks. One was the patriotism of President George W. Bush: "You are with us, or you are with the terrorists," meaning "get on board with plans for war, or ..." Or what? The implication was that real Americans rally around their government and traitors raise critical questions. This poses an obvious problem for journalists, who get paid to raise questions.[13]

Silvio Waisbord questioned the other ways that patriotism could have been manifested that would bolster a critical press during crisis:

Uninterested in questioning the jingoistic drum-banging that took US society by storm after September 11, journalism readily adopted "patriotism as nationalism." Were other versions of patriotism possible? ...Was it possible to understand American patriotism as dissent and freedom of speech, values enshrined in the mythology of US journalism? Could patriotism mean stating that the press freedom was at risk after the Bush administration requested the networks to filter images of Osama bin Laden or announced it would disseminate lies to confound "the enemy"?[14]

Foreign correspondent Chris Hedges said he thinks the role of a patriot is to be a critical dissident and that journalism should not blindly support the government: "We are not cheerleaders. If we want to be popular, we'd better go to another profession ... You need critical distance to have integrity. Unfortunately, there were very, very few people who had that kind of distance."[15] Despite political chants that suggest dissent is patriotic, notions of patriotism are often at odds with one of the key roles of the press, that of the political watchdog.

The Watchdog Role of the Press

Mainstream American journalists who are called terrorists or un-American are paying the price for offering a critical view of the government during a crisis, refusing to neglect their important role as a political watchdog. This watchdog role is threatened when the public, the government, and other journalists pressure the press to demonstrate their patriotism in their news coverage. And what is lost is the crucial function of the press as the Fourth Estate. Television producer Tom Yellin said, "I think the great challenge for the American Press has been to figure out how to cover this in a way that doesn't compromise their journalistic duties but is consistent with what some people view as their patriotic duty. I think that's really been the problem in many instances."[16]

The press is supposed to serve as an overseer of government. Through most of American history, the press as a watchdog was among the nation's most revered principles.[17] The media watchdog role originated with Edmund Burke in seventeenth-century England when he declared that the press had become a Third Estate in Parliament. In the eighteenth century, Cato's Letters, a series of letters written in England by Thomas Gordon and John Trenchard, argued for a free press.[18] Their commentary, which was first printed in the London Journal in the early 1720s, was widely reprinted in the colonies and translated Burke's idea into the fourth branch of government in the soon to be United States, with the executive, legislative, and judicial serving as the other three. During the post–Revolutionary War period in the United States, the colonial press was a vigorous watchdog of the partisan political forces and had thus firmly established the watchdog principle in the colonies by the turn of the century.[19] During the Penny Press era (1830s–1870s), newspapers moved the watchdog role beyond governmental investigations to stories that uncovered excesses of power by key nongovernmental institutions in society.

In the 1890s, yellow journalism brought sensationalism to the news and covered important stories exposing government and corporate corruption.[20] Theodore Roosevelt called the band of investigative journalists in the late 1800s and early 1900s who exposed wrongdoing in society muckrakers, after "Man with the Muckrake," a character in Bunyan's *Pilgrim's Progress*.[21] Muckraking took the form of magazines and books with stories meant to inspire action against excesses and corruption at all levels of power.[22] These investigative journalists played a key role in exposing governmental corruption.[23] Press history reveals a long tradition of a U.S. press that scrutinizes the government, uncovers wrongdoing, and challenges official corruption on behalf of the public—the epitome of the Fourth Estate moniker.

In the modern sense, a watchdog press is one that questions elite sources, filters through the PR machines, and brings issues and information to the people. This may mean questioning those in power, or reporting on stories that hurt the parent company. Independent national security reporter James Bamford said he is able to do critical reports because he is an independent journalist but that those working for big news organizations are not as lucky:

> The problem is that you've got an infrastructure there that is worried about losing viewers or readers and worried about being accused of being unpatriotic, and, as a result, they're putting pressure on the people writing the stories. Subtle and not so subtle benefits come to the people who are keeping the status quo, not to the people who are trying to prove that the president was involved in Watergate or something.[24]

In conflicts such as war or terrorism, journalists face an inner dilemma, which manifests itself in the conflict between a professional responsibility to report the truth no matter how it reflects on your country, and a citizen's national allegiance that makes it hard to abandon responsibility to the homeland. Journalist Ramindar Singh said of the U.S. press after the September 11 attacks that the press

> did not feel compelled to go beyond the surface events that were unfolding. Any criticism of the press' refusal or hesitation to question the administration brought

forth an angry retort from journalists: "You don't understand we are at war," or "this is war we are talking about," suggesting as it were that in a warlike situation the American Press take on a patriotic duty which overrides and supercedes professional duties.[25]

Many media critics argue that today's large news outlets need to adopt a more aggressive watchdog stance, particularly when it comes to covering government policy in times of crisis.[26] Some suggest that elite members of the media seem more interested in cultivating Washington connections than in maintaining skeptical, independent, and not so close relationships with those in power, which would mean challenging the powerful and holding them accountable when they are wrong.[27] Of course, many journalists face an untenable choice because if they challenge the elite connections they often lose access to the information needed to do their jobs. One scholar noted, "Post-September 11 patriotic journalism confirmed the adage that the media want to be loved more than believed."[28]

As Russ Baker wrote in a *Columbia Journalism Review* column in the summer of 2002,

> The need for tough-minded reporting has never been clearer. When journalists hold themselves back—in deference to their own emotions or to the sensitivities of the audience or through timidity in the face of government pressure—America is weakened. Journalism has no more important service to perform than to ask tough, even unpopular questions when our government wages war.[29]

Journalist Dahr Jamail put it this way:

> The two-fold cancer afflicting the majority of establishment media journalists is state pressure and corporate conglomerate/ownership of the media. These now work in tandem, along with the myths of "professional" and "objective" journalism taught in most journalism schools in the U.S. to create the insidious climate of self-censorship. Where critical journalism and real investigations have been replaced by obedient "patriotism" and flag lapels on the black suits of pro-government pundits posing as news anchors. True journalism has no room for patriotism. The master of a true journalist is truth, not the state.[30]

Media outlets must also consider the commercial impact of appearing unpatriotic as Channel 12 on Long Island discovered when management issued a no flag policy for newscasters. Senior vice president of the Cablevision Systems Corporation in charge of News 12 Networks Patrick Dolan said of the policy, "We don't want anyone to get the false impression that our patriotic emotions cloud our reporting of the facts." Advertisers saw the move as unpatriotic and called the station to say so. The *New York Times* reported:

> The businesses, most of them small, that had advertised without incident on News 12 for years—local retailers and Main Street law firms, auto dealers and medical practices—suddenly found themselves perplexed at the imposition of a policy that seemed to defend a journalistic principle few understood in the face of the wave of patriotism that seemed to engulf all of America after Sept. 11.[31]

Journalists must also consider how the audience will respond to critical questions during a crisis. White House reporter Helen Thomas argues that asking pointed questions is often unpopular with the audience and the administration. "From 9/11 on, the American press suddenly had to be super-patriotic or patriots anyway, and they were afraid of being called un-American, unpatriotic..."[32]

Even though Americans are often critical of the way news organizations do their jobs, public support for the news media's role as a political watchdog remains. In every Pew survey conducted since 1985, a majority has said that press criticism of political leaders does more good than harm.[33] However, partisanship plays a role in how much the public supports the watchdog role, with people preferring the watchdog role when the president is not from their political party. For example, Pew research showed that under the presidencies of Ronald Reagan and George H.W. Bush, Democrats were more firmly supportive than Republicans of the role of a watchdog press. When Clinton was in office, Republicans thought that press criticism of political leaders was a good thing more often than Democrats did.[34] The common understanding is that the watchdog should only watch over the political enemy, not the favorite politician.

The magnitude of that partisan divide has grown to record levels during George W. Bush's time in office, with the share of Democrats who believe that press criticism of political leaders "keeps them from doing wrong" increasing throughout Bush's two terms. In 2007, just 44 percent of Republicans believed press criticism of leaders does more good than harm—far lower than the share of Republicans holding this view under the Reagan (65 percent) and Bush Sr. (63 percent) presidencies.[35]

Despite the partisan differences, survey results show that immediately after a crisis, the public expresses support for the press, but soon sees the press as acting too harshly and abusing its power when generally the press is doing nothing more than serving in its watchdog role. A 2001 Pew survey conducted after the September 11 attacks indicated that only 4 percent of Americans stated that the freedom of the press to report what it thought was in the national interest was equal to the government's ability to censor the news.

A few years later, Pew found that Americans actually wanted patriotic news coverage. A Pew poll from February 2003 showed that seven in ten Americans thought it good for coverage to have "a strong pro-American point of view"; however, the same poll reported that a majority of viewers said they valued neutrality in the media. One explanation for this polling result is that the public didn't understand the concept of objectivity or had mixed feelings about it. Or, this poll may suggest that the public didn't include patriotism in their assessment of what neutrality/objectivity means. It may be that the public operates under the assumption that all reports have and should have an American slant. By 2005, 68 percent of Americans wanted neutral rather than pro-American terrorism coverage. However, this, too, is partisan, with 39 percent of Republicans favoring pro-American coverage.[36]

The patriotic response of the press to the September 11 attacks seemed to offer a short-lived legitimacy and popularity to the news media. One Pew research study showed the public rated the news media at an all-time high level in November 2001 for "standing up for America" (69 percent); that figure plummeted twenty points by the summer of 2002.[37] In 2005, the popularity of the press declined further, and a majority once again reported that they believed news organizations do not care about the people they report on. Despite the large number of people who viewed the press as compassionate in November 2001, two months after the attacks, the press did not continue to rate highly on compassion nearly four years later. This shift in the public's attitude corresponds with a change in the news media's coverage about terrorism. The farther from the crisis, the more aggressive the press will become in taking a critical eye to the government, which appears to correspond with the public's attitude about the news media's level of compassion and, subsequently, patriotism.

The trend is similar for the public's assessment of the news media's morality, fairness, and accuracy, all of which returned to pre–September 11 levels by 2002. Despite the fact that those polled viewed news organizations as unwilling to admit mistakes, believed they stood in the way of solving society's problems, and were politically biased, people still continued to value the watchdog role that news organizations perform, with a slight bump in the number who believe press scrutiny of political leaders keeps these leaders from doing things they should not (from 54 percent to 59 percent).[38] This sends a confusing message at best. Perhaps the public is unclear about what the desired press scrutiny should actually look like. When the press actually employs this scrutiny, they are seen by the public as bullies. Even though polls suggest that the public wants a critical press, the same polls show that the public also finds the press to be unpatriotic when they act critically. This also appears to be tied to the partisanship issue; people only see the press as bully when their party is in power.

Some scholars suggest that the watchdog role of the press functions well until the nation enters into a crisis, such as the September 11 terrorist attacks. Historian Howard Zinn has noted that the government tends to enact laws that expand and enhance speech and press protections in times of peace and prosperity but that government constricts these same rights during times of crisis.[39] If the public follows the government's lead, then it would expect the press to be patriotic first and journalists second in times of crisis, only critically analyzing government and its actions when the issues don't invoke strong nationalistic feelings among the general population. Content analyses of the September 11 attacks provided some insight into how broadcast media framed the events in terms of war language and posture, as noted in chapter 5, and highlighted the "you are either with us or against us" attitude that many government officials adopted and communicated. One study found that during September 11 coverage, the media denied the audience a full range of views about terrorism and violated journalistic norms by exhibiting patriotism based on hegemonic assumptions, which did not adequately serve the public good.[40]

Hegemony

Hegemony refers to the success of the dominant classes in presenting their own definition of reality—their worldview—in such a way that it is accepted as common sense and anyone who presents an alternative view is marginalized.[41] The media are seen as controlled by that dominant class and presenting coverage that reflects dominate class interests. Media scholar David Altheide suggests that three testable assumptions pervade the writings on media hegemony.[42] The first connects to the area of media sociology and holds that the socialization of journalists involves general orientations, work routines, and guidelines that are informed by the dominant ideology. The second suggests that journalists tend to cover and frame topics in ways that support the status quo. The third is that U.S. journalists present a pro-American perspective when covering international news, especially when the news involves a third world nation. Altheide cites a number of research studies focused on international news coverage that he says cast doubt on these three assumptions. Yet other researchers who attempted to find research that successfully tested these basic assumptions of media hegemony did find some limited support for the first two assumptions and not the third.[43]

Other researchers have observed hegemonic assumptions of the press during crisis in the United States, including an "us versus them" mentality. During televised coverage of Hurricane Katrina, the hegemonic frame manifested in a few forms: people of New Orleans versus nature during the early coverage, looters versus law enforcement midway through, and then people of New Orleans versus the government during the latter part.[44] Often the hegemonic frame plays on existing stereotypes, as was found in a comparison of newspaper coverage of Hurricanes Andrew in 1992 and Katrina in 2005,[45] and generally in the Katrina coverage with African Americans portrayed as outlaws and savages throughout the coverage.[46] In the initial Oklahoma City bombing coverage that aired on CNN, hegemony allowed the rumor of two Middle Eastern men in a pickup truck as perpetrators to continue on even after it was proven false. In its coverage of the Oklahoma City bombing, CNN noted early in its reporting that the event had international implications, suggesting that foreign hands might be to blame.[47]

A common theme in the news coverage of domestic terrorism is fear of the other. The hegemonic frames of pro-American unity and militarism started early in the coverage of the September 11 terrorist attacks on CNN, when sources asked to make sense of the attacks and repeatedly connected the keyword "America" with freedom and other American ideals, which included the implied notion that it would be un-American to voice political dissent given the magnitude of the day's events. This, of course, runs contrary to the well-known First Amendment values and ideals, yet it was clearly tied to notions of what it means to be a "good" American. By this definition, good Americans stand united and always support the views of the president in times of crisis. The statement by former FBI assistant director James Kallstrom illustrates this and helps to build on the theme of American unity that emerged early in CNN's coverage of the September 11 attacks: "People that hate us and hate what we stand for and hate our way of life have demonstrated that over and over again ... and today they've brought that terrible hatred to the United States of America and we, as a country, as a nation, need to stand together."[48]

America and notions of American values were also employed early in CNN's coverage of September 11, during which official sources began to suggest what would be the appropriate response to the terrorist attacks. For example, at a press conference New York Governor George Pataki observed, "Clearly this is an attack upon America; it's an attack upon our freedom and our way of life, and we must retaliate and go after those who perpetuated this heinous crime against the people of America."[49] In an unrelated press conference several hours later, Bush advisor Karen Hughes delivered a prepared statement that provided a much more subtle example of the use of America in connection with retaliation predicated by an assault on American values: "Our fellow citizens and our freedom came under attack today, and no one should doubt America's resolve."[50]

In examining CNN's initial coverage of the September 11 terrorist attacks, the sources interviewed showed the significance and importance of the concept of American unity—that the press made it seem that the Democrats and Republicans made up the spectrum of all available viewpoints. Once this was established, clear-cut agreement by the two parties suggested that the entire country would be unified regarding all issues related to terrorism. For example, not once during the first twelve hours of coverage did CNN interview any official source with a political affiliation other than Democrat or Republican. Not once during the first twelve hours did anyone, source or journalist, suggest that an option other than supporting the president would exist. During a crisis, the press is clearly hesitant to contradict the so-called party line, in this case the line was a bipartisan, progovernment line.

New York's Democratic Senator Hillary Clinton, typically at odds with the Bush administration's agenda, clearly stated her support for him in the context of the need for American unity during times of crisis: "We're all united behind the president ... the legislative leadership of our country, from both Houses of Congress, on both sides of the aisle, saying just as clearly as we can that this was an attack on America. And the president of the United States is our president. And we will support him in whatever steps he deems necessary to take..."

And this idea of unity was also reinforced by journalists, as this observation by CNN correspondent Jamie McIntyre makes clear: "[General Hugh Shelton, Chairman of the Joint Chiefs of Staff] was flanked with bipartisan support from Capitol Hill. The leading Democrats and Republicans from the Senate Armed Services Committee there to show support that the United States, when attacked, has no divisions in its political will."

It is noteworthy that unity was defined in such a narrow way. CNN gave viewers the perception that because government leaders from the two dominant American political parties were unified, the entire country was unified. While this most likely accurately reflected a majority view, it discounted many other available viewpoints that in theory might have altered, or at least encouraged, some substantive issue debate. Further, it emphasized the patriotic and therefore "government deemed appropriate" response to the terrorist attack. Nichols and McChesney noted that the September 11 coverage was at times dazzling and that it did appear to bring the country together:

> But the immediacy and drama of the coverage of the 9/11 attacks and the immediately ordained war on terror also meant that elementary questioning and investigation of Bush administration conduct and policy initiatives were largely missing. In a climate where the political Right was bloviating on the flames of blind nationalism, to even raise a tough question was unpatriotic, if not treasonous. [51]

Early on in the coverage, the hegemonic statements came from government officials, but it was not long before the media started to incorporate hegemonic language into their own coverage, adapting the inevitability of a military response. It is possible that a journalist's individual ideology could trump traditional journalistic norms and routines during national crisis. This appeared to be the case for many reporters during the September 11 attacks when a string of highly unusual statements from journalists were aired on national television. Former NBC anchor Tom Brokaw was unabashed in displaying his nationalism on air the morning of September 11, making the following statements during the first five hours of NBC News coverage:

> This country, the most powerful in the world, has been, in effect, semi-immobilized today by these terrorist attacks, and we can only hope that they are now all over with.

> There's a picture of Lower Manhattan, ladies and gentlemen, the most important city in the world in so many ways, and now it has been attacked by terrorists at the World Trade Center, and the damage is beyond our ability to tell you in great detail.

> It's—it is difficult to comprehend, that this country, the strongest country in the world, has been the target of a major coordinated terrorist attack.

> A good portion of the nation's capital, the most powerful center in the world, has
> been immobilized as well, as a result of these terrorist acts.

Brokaw's nationalistic bias may have been a product of reporting on a national crisis when you feel personally attacked, a phenomenon discussed in chapter 6. But these statements also suggest an alliance with the government to propagate the administration's statements about the attacks. President Bush suggested that the attacks had no underlying motive, other than a jealousy about what is good in America and a hatred of freedom. Nicholas and McChesney likened the press coverage after 9/11 to what you expect in a "kept press system in an authoritarian regime."[52] For example, former CBS anchor Dan Rather was explicit in his support for the government when he said on air:

> Let no one, no matter how you feel politically, misunderstand: The president of the
> United States is in charge. He is the commander in chief. And whatever anybody
> thinks of him, or doesn't think of him at any moment, he's the commander in chief.
> He's the leader of the nation. He is on Air Force One. He is in—in instantaneous
> communication when he wants to be anyplace in the world, and there should not be
> any misunderstanding about that. There'll be plenty of time later in the days and
> weeks to follow about whether he and his White House operation deported them-
> selves as they could have and should have today. But as of this moment, no one
> should have any—any misunderstanding about the president being in charge.

Comments from Rather and Brokaw are particularly worthy to note because they come from news anchors who may see their journalistic role differently. Instead of serving as traditional watchdogs, anchors may sometimes feel the need to serve as leaders, offering a calming presence, serving as a primary and consistent voice for their network and their audience during a moment of national crisis, fear, and confusion. The term "anchor" as used in television news was coined by legendary CBS producer Don Hewitt. He used the term anchor to signify the way the lead person on air in a newscast anchors the show. He likened the television anchor to the anchor leg of a relay race. Both of these definitions hint at what is expected of the anchor—stability and resolve to complete the task.

Washington Post columnist Marc Fisher wrote of former ABC anchor Peter Jennings' sixty hours of desk time the week of the September 11 attacks:

> We watched Peter Jennings' beard grow, and we were somehow reassured that he did
> not shave, that through morning, afternoon, evening and on into the night, he did

not leave the desk, that he confided in us his uncertainties, that he shared the confusions of each hour. He grew more pale and more vulnerable, as if he knew that we needed him to be human, so that we could be together.[53]

Jennings's normal restraint was still there, but he offered advice, something Jennings himself noted that anchors do not normally do:

We do not very often make recommendations for people's behavior from this chair but as [one ABC News correspondent] was talking, I checked in with my children, and it, who were deeply stressed, as I think young people are across the United States. So, if you're a parent, you've got a kid in some other part of the country, call them up. Exchange observations.[54]

During a live forum with the BBC, Jennings talked about how his extensive experience with reporting crisis helped, but did not ensure flawless crisis reporting:

It's hard to stay calm on air because I think that, like anybody else, we're having a tremendously, tumultuous emotional experience. But to be honest I think it is training. I've been on the air for lots of disasters—none as severe as this. I was on the air for 11 hours at the time of the Challenger explosion and I have been on the air through natural disasters, political disasters and assassinations. However, you're so busy and you're so focused on trying to bring everybody together and make the whole thing comprehensible that it's perhaps easier for the anchor persons to stay calm at that moment than it were if I were simply watching television like everybody else.[55]

Patriotic Journalism

Patriotic journalism is said to be dangerous because it "denies to the public the information and detached perspective people need to make sound decisions."[56] Yet after September 11, evidence of patriotic journalism abounded. The broadcast and cable news networks in the United States consistently supported the Bush administration immediately following the 9/11 attacks, and public opinion polls show that this is what a majority of the public expected from the media. Some research has shown that a lack of diverse sources, a lack of differing viewpoints, and the use of headlines and graphics that said "America under Attack" combined to create early news frames that suggested that a military response to the attacks was justified and that America was unified after the attacks.[57]

Statements made on air during the initial live television coverage on September 11 demonstrate how quickly journalists assumed the role of the patriot. But, more subtle language such as the use of personal pronouns also illustrated the collective nationalistic mentality of journalists at the time. Journalists used "we" to refer to the United States and "they" to refer to anyone else. The coverage was named in the same manner, with titles such as "A Nation Responds" suggesting the whole nation was delivering a unified response. The coverage supported this approach. Dissenting opinions were not heard on air for some time after September 11. From the media coverage of the September 11 attacks, we can conclude that the country was of one mind on the cause of the attacks and the appropriate response. Even when journalists interviewed politicians from both sides of the aisle, they all spoke of a unified country and a unified response of retaliation. While this certainly reflected the majority view at the time, it didn't reflect the full scope of viewpoints and beliefs. Talk shows decided to forgo their opening monologues or relied on anemic humor. Bill Maher's *Politically Incorrect* talk show offered a contrary view, which caused massive advertiser dropout and the show was eventually cancelled, although ABC said this was because of low ratings rather than Maher's statements.[58] When panelist Dinesh D'Souza said that he didn't think the terrorists were "cowards," as President Bush had described them, host Bill Maher said, "We have been the cowards. Lobbing cruise missiles from two thousand miles away. That's cowardly. Staying in the airplane when it hits the building. Say what you want about it. Not cowardly. You're right."

White house press secretary Ari Fleischer responded to a press conference question about Maher's comments by saying that Americans should "watch what they say. This is not a time for remarks like that. There never is."[59] Maher actually began that September 17 show with a statement about the importance of a critical media during a crisis:

> I do not relinquish—nor should any of you—the right to criticize, even as we support, our government. This is still a democracy and they're still politicians, so we need to let our government know that we can't afford a lot of things that we used to be able to afford. Like a missile shield that will never work for an enemy that doesn't exist. We can't afford to be fighting wrong and silly wars. The cold war. The drug war. The culture war.[60]

Maher later wrote a piece in the *Los Angeles Times* about that period:

But the atmosphere in the fall of 2001 allowed for very little beyond singing "God Bless America" and buying a flag to put on your gas-guzzling, terrorist-funding car. In fact, I was not the only one whose comments helped flood the ABC switchboard in the first day or two after the attacks. Peter Jennings had the temerity to suggest that some presidents are more reassuring than others in situations of crisis. Stone him! Kill him! How dare he suggest that our president might be anything but perfect at everything! We have been attacked; ipso facto, our president is a genius![61]

Some experts suggest that the push for patriotism was evident in the days following the initial 9/11 coverage when stories that conflicted with the dominant governmental view were not written, and the way the Bush administration framed the attacks went unchallenged. The media lens that should have been focused on the president and the government was clouded by a clear push to patriotism. Scholars have noted this "rally around the flag" effect in the past, where during times of national crisis the press and public both become more favorable to the nation's leaders. When terrorism hits home, the press also feel under attack by the terrorists and in turn become less critical and more assuring, because this is the patriotic thing to do. Polls show that this is what the majority of the public want and the press is in line with the majority view so it sees no problem with a nationalistic approach.

Scholars, activists, and some politicians could have and would have offered dissenting opinions, but they were either silent or not offered a chance. Howard Zinn spoke in October 2001 at Elon University in North Carolina. He offered a different analysis of September 11 than the one offered in the mainstream media. When asked why no one was interviewing him, he said clearly this view was not wanted in the mainstream press at this time, and the suggestion was that it was not wanted by neither the press nor the public. Zinn proffered stories about acclaimed writers trying to find publication venues for pointed essays about the attacks. The media self-censorship is one of the more troubling aspects of media professionals blinded by patriotism during a national crisis because many important stories are never covered and the public is none the wiser.

Writer Katha Pollitt did find a place for her dissenting ideas three weeks after September 11 when she posted on The Nation Web site a column that was also published in the October 8, 2001, print issue. She discussed, among other issues, the debate she and her daughter were having about whether to fly the flag out their window. "Definitely not, I say: The flag stands for jingoism and vengeance and war. She tells me I'm wrong—the flag means standing together and honoring the dead and saying no to terrorism. In a way we're both right…" This column provoked a spiteful uproar. "Pollitt, honey, it's time to take your brain to the dry cleaners," one columnist wrote; "We're at war, sweetheart," wrote another. Bernard Goldberg named her seventy-four on his list of 100 People Who Are Screwing Up America book.

The networks also gussied up the visual images in the news coverage after 9/11 to portray more patriotism. American flag images and colors popped up everywhere. The banner on CNN was red, white, and blue and waved like a flag. News anchors wore flag pins. Local newscasts aired promos with statements about the staff supporting the troops.

In the wake of the attacks, media critics began to challenge those journalists and news organizations who were wearing their patriotism in their professional lives in the form of red, white, and blue ribbons and flag pins on clothes and bold patriotic graphics on air. On September 20, 2001, the New York Times published an article about the majority of cable news shows using the flag in their broadcasts. In that article, Alex Jones—director of Harvard's Joan Shorenstein Center on the Press, Politics, and Public Policy—said that immediately after 9/11 the emotional, editorialized, supportive coverage was understandable, but now the news must remain vigilant and critical so as to remain loyal to the criteria of their jobs.[62] A few months later, Ruth Conniff complained in the Progressive that it is "sad that a lot of journalists have little to do but wear flag lapel pins and read Pentagon press releases about the war…"[63] She also criticized the cable networks for creating catchy phrases such as "America's New War" and playing "drum-and-horn" music that made it seem as if the networks were operating as an arm of the government and marching off to war with them.[64]

The Los Angeles Times quoted Jimmy Kelly of Time magazine as saying, "our friends in cable TV… have muddied the waters… [T]hey plant the flag on their screens and try to stick a waving flag on virtually everything that moves, and the subtle implication in that the network has gone to war as well, on the side of the U.S. troops."[65] Prior to September 11, scholars had explored the ways that "flag symbolism is almost always associated with a sustained propaganda program … and it is patently clear that certain groups claim a monopoly over the flag and attempt to exclude those who do not accept their definition of patriotic loyalty."[66] One commentator suggested that "[i]f any of the pillars of journalism have been shaken [since 9/11] it has been independence."[67]

And yet, when media organizations tried to be more neutral in the coverage, this brought harsh criticism from conservative media watchdog groups. These watchdog groups cheered ABC's Cokie Roberts for her decision to get around the policy by wearing an eagle lapel pin instead. Network policies, which restricted staffers from wearing flag pins, were reviled in the conservative press. ABC staffer Ted Koppel at a Brookings Institution forum broadcast by C-SPAN said, "I don't believe that I'm being a particularly patriotic American by slapping a little flag in my lapel and then saying anything that is said by any member of the U.S. government is going to get on without comment and anything that is said by someone from the enemy, is immediately going to be put through a meat grinder of analysis. Our job is to put it all through the meat grinder of analysis." The ABC policy and Koppel's remarks were lambasted by conservative watchdog groups, affirming the previous surveys that showed that audience response is quite partisan.[68]

Meanwhile, the following exchange with Howard Kurtz and CBS anchor Dan Rather on the CNN program *Reliable Sources* prompted Rather bashing from both conservative and liberal watchdog groups—conservatives because he was not wearing the flag, and liberals because he too emphatically highlighted his own patriotism:

Kurtz: Well, speaking of patriotic Americans, there is a bubbling controversy in the business, as you probably know, about whether journalists on the air should wear these little lapel flags. NBC's Tim Russert did it on *Meet the Press*. ABC News has barred its people from doing that. Does it seem to you that journalists who show the flag are being patriotic, or are they somehow kind of turning it to cheerleaders for team USA?

Rather: I have no argument with anyone who does it, but I don't because it doesn't feel right to me. I have the flag burned in my heart, and I have ever since infancy. And I just don't feel the need to do it. It just doesn't feel right to me. And I try to be, particularly in times such as these, and I have tried to be in touch with my inner self, my true inner self, and I tried to listen. And my inner self says you don't need to do that. But I have absolutely no argument with anyone else who feels differently."[69]

Rather was also not safe from liberal criticism either after he ended a special hour-long version of the CBS Nightly news with this patriotic speech:

After a second wave of U.S. aerial bombardment in the Afghan night, the America mind runs on two tracks: First, our thoughts and our love are with our warrior men and women, our sons and our daughters, brothers and sisters, husbands and wives, fathers and mothers over there. We see them at work and we are reassured by their professionalism and skill. We know some may come back in flag-draped caskets, but we reluctantly and sadly accept that as a reality of a war forced upon us. Less familiar to Americans are concerns of protecting home and hearth here. Our worst fears tell us that the enemy within could strike again. On a day when President Bush swore-in the first Director of Homeland Security in our history, thoughts turn to the words of the late newsman Elmer Davis. "This will remain the land of the free only so long as it is the home of the brave."[70]

Shortly before the start of the Iraq War, PBS's Bill Moyers donned the flag pin to counter those he claimed had hijacked it: "When I see flags sprouting on official lapels, I think of the time in China when I saw Mao's little Red Book on every official's desk, omnipresent and unread." The argument surrounding the wearing of the flag actually hinged on the perceived meaning of the flag, supporting the country, supporting the troops, or advocating for the war.

CNN's Lou Dobbs, whose show donned intense patriotic imagery, defended the flag images in his weekly commentary on October 10, 2007, and criticized fellow journalists who saw it differently:

Katie Couric of CBS News takes exception to "the whole culture of wearing flags on our lapel and saying 'we' when referring to the United States." We are Americans, right, Katie? I'm sorry, but how can anyone possibly be offended by acknowledging that our troops who are sacrificing so much for us are ours, and that we are their proud countrymen?

Dobbs also addressed critics who suggested that wearing a flag pin can undermine objectivity: "I've never been disinterested when it comes to this country, and I'm certainly not disinterested about the welfare of our troops nor the outcome of conflicts involving this nation."

Audience Impact

Despite the discussion among critics and journalists, the audience does not even seem to evaluate stories differently based on use of patriotic images. A study that examined the effects of such patriotism and nationalism on news audiences found that across the sample, the audience did not respond differently to stories with patriotic images.[71] This study showed that viewers did not perceive journalistic norms emphasizing fairness, newsworthiness, avoiding bias, accuracy, importance, and balance to be compromised by patriotic images in news stories. However, this study showed that Democrats and liberals who saw a patriotic version of a story were significantly more likely to perceive the anchor as less objective than Democrats and liberals who saw the neutral version.[72] This suggests perhaps that the anchor perception does not compromise perceived quality and objectivity of the news story. And perhaps this finding is something that the business side of news already knows. Patriotism in the news is a by-product of the commercial nature of the news business today. Seib notes that "Fox, in particular, staked claim to a niche audience of viewers with a fondness for red-white-and-blue coverage, and when it was successful imitators followed."[73] The networks obtained large audiences when they didn't present negative reporting or stories contrary to the administration's line because that was what the public wanted to hear.[74]

The audience response was clearly on the minds of everyone two months after September 11, when then National Security Advisor Condoleezza Rice teleconferenced with six television news executives and asked them to limit the use of a videotaped address by al Qaeda leader Osama bin Laden. She argued that the tape might have a negative effect on the public and also that it might contain a coded message. These statements were widely available on the Internet and satellite, but the TV news executives agreed to limit their airtime. When the next tape arrived, MSNBC and Fox News did not air it, and CNN showed only brief excerpts. Because these videos were also readily available on the Internet, it seems the government and the press both wanted to limited play of these tapes for reasons other than national security, to prohibit any delay in the march to war in the case of the government, and to prohibit an angry audience in the case of the networks.

The government also asked Voice of America radio not to air an interview with Mullah Mohammed Omar, the leader of Afghanistan's ruling Taliban. In this case the government even claimed that this interview was not newsworthy, a ludicrous claim under any definition of newsworthiness. When the VOA staff complained, the government removed the request and the interview was aired.

In June 2006, the *New York Times* published two reports about secret antiterrorism programs being run by the Bush administration. Critics claimed that the paper was being unpatriotic, was aiding the terrorists and should be indicted under the Espionage Act. Stephen Spuriell wrote for the *National Review* online on June 22, 2006: "Keller and his team really do believe they are above the law. When it comes to national security, it isn't the government that should decide when secrecy is essential to a program's effectiveness. It is the

New York Times. National security be damned. There are Pulitzers to be won."

Blogger Bryan Preston wrote that same day: "[T]he NY Times outing a legal program organized to stop terrorists, and the only beneficiaries of outing this program turn out to be the terrorists who are now warned and the reporters who stand to gain from book deals, Pulitzers and the adulation of their increasingly disgusting peers."

The Big Story with John Gibson on the Fox News Network opened the June 23, 2006, show with this statement from host John Gibson: "The New York Times has outed another government anti-terrorism program, one that kept tabs on banks and terror financing and had apparently been working out pretty well. But that didn't stop the 'paper of record' and some other newspapers from letting everybody know about it."

Representative Peter King urged the Bush administration to prosecute the paper. "We're at war, and for the Times to release information about secret operations and methods is treasonous," the New York Republican told the Associated Press. King charged that the paper was "more concerned about a left-wing elitist agenda than it is about the security of the American people." Times Watch from the conservative media watchdog group Media Research Center asked, "Exactly what 'public interest' is being served by making it that much harder to fight terrorism?"

In an editorial, the Times defended its decision and cited other instances when the newspaper did not print information when the stories were deemed too dangerous to print: such as when the newspaper did not print the Kennedy administration's plans for the Bay of Pigs invasion, and when the paper withheld a story on the Bush administration's secret antiterror wiretapping program for more than a year while listening to governmental objections. The Times editorial suggested that these decisions are not made lightly and that the press used good judgment when deciding to print, concluding that "The free press has a central place in the Constitution because it can provide information the public needs to make things right again. Even if it runs the risk of being labeled unpatriotic in the process."[75]

Some letters to the editor in response to the editorial were supportive:[76]

Henry A. Lowenstein, New York, June 28, 2006: It is a sad day when The Times feels the need to use prime editorial space to defend itself from doing what the 'paper of record' is supposed to do: report the news accurately, truthfully and without bias. I am proud to subscribe to The Times and will continue to write letters supporting and opposing the views stated on your editorial page. That is what our bravest men and women are dying for in Iraq, isn't it?

Others criticized the decision to run the story:

Bill Decker, San Diego, June 28, 2006: With your self-imposed mandate to "provide information the public needs," The Times would do well to remember that the current enemy of our country is not fighting us as a uniformed army but as part of that public. Do you really intend to fulfill their information needs as part of your mandate?

Monte Seewald, Old Bridge, N.J., June 28, 2006: Your editorial "Patriotism and the Press" brings into play the concept of "could have" versus "should have." Could you print the story of the secret antiterrorism program in question? Yes. Should you have? No.

Some of these examples suggest that during conflict or crisis media are more likely to be afraid to dissent from majority ideas or to report dissention,[77] in part because they're afraid to look unpatriotic. They may also fear a legal battle.[78] As previously noted, surveys indicate that this fear is well-founded—public opinion polls show that the public welcomes government censorship during times of crisis. Further, those who challenge patriotic news are sometimes attacked for being unpatriotic themselves. Dan Rather raised this issue quite prominently, expressing fear that the American press would "be more docile," in part "out of fear that you'd be called a bad name, unpatriotic, or radical right or liberal."[79] His concern was not unrealistic when one notes just a few incidents that received media coverage: the increased pressure on political cartoonists,[80] the cancellation of Phil Donahue's show,[81] and the labeling of CNN as unpatriotic when it aired a videotape of snipers targeting U.S. troops.[82] NBC News President Steve Capus said that "you have the Pentagon flaying the media, saying it's unpatriotic to ask questions."[83] As former Senator Alan Simpson put the question on the News Hour on November 6, 2001, "Are you a patriot first or are you someone who remains neutral?"

Others saw this issue in less dichotomous terms. Nelson Poynter Scholar for Journalism Values Bob Steele suggested on a September 20, 2001, Web entry that "the true measure of journalism's worth to our democracy will be measured not by our outward displays of patriotism, but by the work we produce ... by the vigor and rigor we bring to our coverage and commentary."

Steele revisited this subject on the following fourth of July and asked:

for those of us who practice journalism in this democracy of the United States of America, can we show support for our country while also offering news coverage that reveals the weaknesses of our system? While offering commentary that challenges our leaders and their policies? Can journalists be loyal to this country while provoking spirited debate that might lead to dissension?

These questions are addressed with starkly different answers depending on the political proclivities of the journalist, watchdog group, or government official who answers them. The public does not appear to be bothered by patriotic images, and surveys show that they expect a pro-American slant to coverage. But journalists who report from a variety of perspectives, including dissenting ones, are routinely accused of anti-American behavior. The journalists accused of being opinionated have addressed the un-American label head on. In an April 1996 commentary in *Quill*, CNN's Christiane Amanpour explained how she defined objectivity:

> The very notion of objectivity in journalism becomes immensely important ... I have come to believe that objectivity means giving all sides a fair hearing, but not treating all sides equally. Once you treat all sides the same in a case such as Bosnia, you are drawing a moral equivalence between victim and aggressor. And from here it is a short step to being neutral. And from there it's an even shorter step to becoming an accessory to all manners of evil ... Objectivity must go hand in hand with morality.[84]

Amanpour's brand of objectivity seems unpalatable to the American people. Six years after 9/11, almost half of U.S. news consumers still find fault with the U.S. media for failure to stand up for America with 43 percent of the general public criticizing the press for being too critical of America. The younger and more educated group of people whose main source of news is the Internet are particularly likely to criticize news organizations for their failure to "stand up for America." An Internet News Audience survey in 2007 showed that 53 percent of those who get most of their news from the Internet believe that news organizations are too critical of America, compared to 38 percent of those who get their news mainly from TV and 42 percent who get there news mainly from newspapers.

Considering these numbers and previous public opinion polls that show people think news organizations are too critical of America, Amanpour's definition of objectivity reflects a dedication to the public interest better suited for an international audience. When first director general of the BBC John Reith rejected the idea that it is the responsibility of the broadcaster to give the public what they want, he said:

> Public service is the maintenance of high standards, the provision of the best and the rejection of the hurtful … it is occasionally indicated to us that we are apparently setting out to give the public what we think they need—and not what they want—but few know what they want and very few what they need—in any case it is better to overestimate the mentality of the public than to underestimate it.[85]

In America, broadcasters disagreed with Reith's public service model. In 1933, the National Association of Broadcasters produced a pamphlet that asserted:

> The nervously active American is never in a mood to take educational punishment. You must interest him—or he quickly tunes you out. This characteristic is in only slightly lesser degree fundamental to any discussion of listener reaction in any country. It is the rule and the law and the testament upon which every successful broadcast structure is based. It is the risk, for instance, that Sir John Reith runs in Britain when he avowedly gives his public what he believes it is good for it to have.[86]

Public opinion polls offer a confusing message: people want a pro-American, watchdog press that does not go too far and can be censored by the government, which is an unfortunately untenable combination of factors. Perhaps terrorism coverage at its best gives the audience what it needs versus what it wants, something the U.S. press has been reluctant to do.

Notes

1   Ariel Merari, "Terrorism as a Strategy of Insurgency," in Gérard Chaliand and Arnaud Blin, eds. *The History of Terrorism: From Antiquity to Al Qaeda* (Berkeley: University of California Press, 2007), 13.

2   Pew Research Center for the People and the Press, "Who Flies the Flag? Not Always Who You Might Think: A Closer Look at Patriotism," June 27, 2007, http://pewresearch.org/pubs/525/who-flies-the-flag-not-always-who-you-might-think.

3   Tom W. Smith and Lars Jarkko, *National Pride in Cross-national Perspective* (Chicago, Ill.: University of Chicago National Opinion Research Center, 2001). http://209.85.207.104/search?q=cache:i0X_XkGW9s4J:www.issp.org/Documents/natpride.doc+University+of+Chicago+%E2%80%9CNational+Pride:+A+Cross-national+Analysis&hl=en&ct=clnk&cd=1&gl=us&client=firefox-a

4   Robert Jensen, *Citizens of the Empire: The Struggle to Claim our Humanity* (San Francisco: City Lights Books, 2004), 40.

5   Leo Tolstoy, "Patriotism and Government," available online at http://dwardmac.pitzer.edu/Anarchist_Archives/bright/tolstoy/patriotismandgovt.html.

6   Alasdair MacIntyre, "Is Patriotism a Virtue?" in R. Beiner, ed. *Theorizing Citizenship* (New York: State University of New York Press, 1995), 209–228; Joshua Cohen and Martha C. Nussbaum, *For Love of Country: Debating the Limits of Patriotism, Beacon Press* (Boston: Beacon Press, 1996); Jürgen Habermas, "Appendix II: Citizenship and National Identity," in William Rehg, trans. *Between Facts and Norms: Contributions to a Discourse Theory of Law and Democracy* (Cambridge, Mass.: MIT Press, 1996).

7   Martha Nussbaum, "Patriotism and Cosmopolitanism," *Boston Review*, 19, 5 (Fall 1994). http://www.soci.niu.edu/~phildept/Kapitan/nussbaum1.html.

8   Emma Goldman, "Patriotism: A Menace to Liberty," in *Anarchism and Other Essays* (New York: Dover, 1969).

9   Paul Gomberg, "Patriotism Is Like Racism," in Igor Primoratz, ed. *Patriotism* (Amherst, N.Y.: Humanity Books, 2002), 105–112.

10 David McCabe, "Patriotic Gore, Again," in Igor Primoratz, ed. *Patriotism* (Amherst, N.Y.: Humanity Books, 2002), 121–141.

11 James Boswell, "Entry for Friday, April 7, 1775," *Life of Johnson* (New York: Everyman's Library, 1993), 615.

12 Samuel Johnson, *Idler* #30 (November 11, 1758) as quoted on http://www. samueljohnson.com/patrioti.html.

13 Robert Jensen, "Journalism Should Never Yield to 'Patriotism,'" Published on Wednesday, May 29, 2002 in the Long Island, New York *Newsday*. Retrieved from the commondreams.org website http://www.commondreams.org/views02/0529-02.htm.

14 Silvio Waisbord, " Journalism, Risk, and Patriotism," in Barbie Zelizer and Stuart Allan, eds. *Journalism After September 11* (London: Routledge, 2002), 207.

15 Chris Hedges, "We're Not Mother Teresas," in Kristina Borjesson, ed. *Feet to the Fire: The Media after 9/11* (Amherst, N.Y.: Prometheus Books, 2005), 531–532.

16 Tom Yellin, "Inside the Ratings Vise," in Kristina Borjesson, ed. *Feet to the Fire: The Media after 9/11* (Amherst, N.Y.: Prometheus Books, 2005), 49.

17 Leonard W. Levy, *The Emergence of a Free Press* (New York: Oxford University Press, 1985).

18 John Trenchard and Thomas Gordon, eds., *Essays on Liberty, Civil and Religious, and Other Important Subjects* (Indianapolis: Liberty Fund, 1995).

19 David A. Copeland, *The Idea of a Free Press: The Enlightenment and Its Unruly Legacy* (Evanston, Ill.: Northwestern University Press, 2006); Jeffery A. Smith, *Printers and Press Freedom: The Ideology of Early American Journalism* (New York: Oxford University Press, 1988).

20 Sidney Kobre, *The Yellow Press and Gilded Age Journalism* (Tampa, Flo.: Florida State University, 1964).

21 Wayne Andrews, ed., *The Autobiography of Theodore Roosevelt, Condensed from the Original Edition, Supplemented by Letters, Speeches, and Other Writings* (New York: Charles Scribner's Sons, 1913, rep. 1958).

22 Hazel Dicken-Garcia, *Journalistic Standards in Nineteenth-Century America* (Madison, Wis.: University of Wisconsin Press, 1989).

23 Carl Jensen, ed., *Stories That Changed America: Muckrakers of the 20th Century* (New York: Seven Stories Press, 2000).

24 James Bamford, "Dangerous Nonsense," in Kristina Borjesson, ed. *Feet to the Fire: The Media after 9/11* (Amherst, N.Y.: Prometheus Books, 2005), 323.

25 Ramindar Singh, "Covering September 11 and Its Consequences: A Comparative Study of the Press in America, India and Pakistan," in Nancy Palmer, ed. *Terrorism, War and the Press* (Cambridge, Mass.: Joan Shorenstein Center on the Press, Politics and Public Policy, 2003), 43.

26 John Nichols and Robert W. McChesney, *Tragedy and Farce: How the American Media Sell Wars, Spin Elections, and Destroy Democracy* (New York: New Press, 2005).

27 Noam Chomsky, *The Culture of Terrorism* (Boston, Mass.: South End Press, 1988); Alexander Cockburn and Jeffrey St. Clair, *End Times: The Death of the Fourth Estate* (Petrolia, Calif.: Counter Punch, 2007); Helen Thomas, *Watchdogs of Democracy?: The Waning Washington Press Corps and How It Has Failed the Public* (New York: Charles Scribner's Sons, 2006).

28 Waisbord Silvio, "Journalism Risk and Patriotism," in Barbie Zelizer and Stuart Allan, eds. *Journalism After September 11* (London: Routledge, 2002), 201–216.

29 Russ Baker, "Want to Be a Patriot? Do Your Job," *Columbia Journalism Review* (May/June 2002): 41.

30 Dahr Jamail, e-mail message to the author, March 10, 2008.

31 Warren Strugatch, "When Patriotism and Journalism Clash," *New York Times*, October 7, 2001,http://query.nytimes.com/gst/fullpage.html?res=9C0DE2DB163CF934A35753C1A9 679C8B63.

32 Helen Thomas, "Grande Dame: Persona Non Grata," in Kristina Borjesson, ed. *Feet to the Fire: The Media after 9/11* (Amherst, N.Y.: Prometheus Books, 2005), 323.

33 Pew Research Center for the People and the Press, "Public More Critical of Press, But Goodwill Persists: Online Newspaper Readership Countering Print Losses," Released June 26, 2005, http://people-press.org/reports/display.php3?ReportID=248.

34 Pew Research Center for the People and the Press, "Internet News Audience Highly Critical of News Organizations: Views of Press Values and Performance: 1985–2007," Released August 9, 2007. http://people-press.org/reports/display.php3?ReportID=348.

35 Pew Research Center for the People and the Press, Internet News Audience Highly Critical of News Organizations, http://people-press.org/reports/display.php3?ReportID=348.

36 Pew Research, "Public More Critical of Press, But Goodwill Persists."

37 Pew Research Center for the People and the Press, "News Media's Improved Image Proves Short-Lived: The Sagging Stock Market's Big Audience," August 4, 2002, http://people-press.org/reports/display.php3?ReportID=159.

38 Pew Research Center for the People and the Press, "News Media's Improved Image Proves Short-Lived."

39 Howard Zinn, *Declarations of Independence: Cross-Examining American Ideology* (New York: Harper Perennial, 1990).

40 Sandra Borden, "Communitarian Journalism and Flag Displays after September 11: An Ethical Critique," *Journal of Communication Inquiry*, 29, 1 (2005): 30–46.

41 Perry Anderson, "The Antinomies of Antonio Gramsci," *New Left Review*, 100 (1976): 5–78; Walter L. Adamson, *Hegemony and Revolution: A Study of Antonio Gramsci's Political and Cultural Theory* (Berkeley: University of California Press, 1980).

42 David L. Altheide, "Media Hegemony: A Failure of Perspective," *Public Opinion Quarterly*, 48 (1984): 476–490.

43 Werner J. Severin and James W. Tankard, Jr., *Communication Theories: Origins, Methods, and Uses in the Mass Media* (3rd ed.) (New York: Longman, 1992).

44 Kathleen Tierney, Christine Bevc, and Erica Kuligowski, "Metaphors Matter: Disaster Myths, Media Frames, and Their Consequences in Hurricane Katrina," *The ANNALS of the American Academy of Political and Social Science*, 604, 1 (2006): 57–81; Worapron Worawongs, Weirui Wang, and Ashley Sims, "U.S Media Coverage of Natural Disasters: A Framing Analysis of Hurricane Katrina and the Tsunami" (Paper presented at the annual meeting of the Association for Education in Journalism and Mass Communication, August 8, 2007).

45 Caroline Heldman, "'Burnin' and 'Lootin': Race and Media Coverage of Hurricane Katrina" (Paper presented at the annual meeting of the Western Political Science Association, La Riviera Hotel, Las Vegas, Nevada, March 8, 2007).

46 Michael Eric Dyson, *Come Hell or High Water: Hurricane Katrina and the Color of Disaster* (New York: Basic Civitas Books, 2006).

47 Amy Reynolds, "How 'Live' Television Coverage Affects Content: A Proposed Model of Influence and Effects" (Paper presented to the International Communication Association Conference Mass Communication Division, Montreal, Quebec, May 23, 1997).

48 http://transcripts.cnn.com/TRANSCRIPTS/0109/11/bn.46.html.

49 http://transcripts.cnn.com/TRANSCRIPTS/0109/11/bn.42.html.

50 http://www.whitehouse.gov/news/releases/2001/09/20010911-11.html.

51 Nichols and McChesney, 55.

52 Nichols and McChesney, 55.

53 Patricia Sullivan, "ABC News Anchor Was a Voice of the World," *Washington Post*, August 8, 2005, A01.

54 "Peter Jennings Reporting," *New York Observer*. http://gawker.com/news/new-york-observer/peter-jennings-reporting-116654.php.

55 "Out of the ashes—Live from New York: How has New York and America coped with the Tragedy of 11 September?" Public Forum, December 17, 2001, on the BBC Web site, available at http://news.bbc.co.uk/1/hi/in_depth/americas/2001/nyc_out_of_the_ashes/170 5631.stm.

56 Montague Kern, "Introduction," in W.L. Bennett and D. Paletz, eds. *Taken by Storm: The Media, Public Opinion and US Foreign Policy in the Gulf War* (Chicago: University of Chicago Press, 1994).

57 Amy Reynolds and Brooke Barnett, "America under Attack: CNN's Verbal and Visual Framing of September 11th," in Steve Chermak, Frankie Y. Bailey, and Michelle Brown, eds. *Media Representations of September 11th* (New York: Praeger, 2003).

58 Jonah Goldberg, "Maher's Final Half Hour: Why PI Should Go," *National Review Online*, September 28, 2001. http://article.nationalreview.com/?q=ZDc2MjM2ZTFkODAyYWE0OGJk Nzc2NDliNmIzYjVkZGU=; Celestine Bohlen, "Think Tank; in New War on Terrorism, Words Are Weapons, Too," *New York Times*, September 21, 2001; "Maher Tapes Final Episode of 'Politically Incorrect," *USA Today* online, June 29, 2002.

59 Bill Carter and Felicity Barringer, "A Nation Challenged: Speech and Expression; in Patriotic Time, Dissent Is Muted," *New York Times*, September 28, 2001.

60 J. Edward Tremlett, "'Politically Incorrect' Cancelled—Was it ABC or APC?" *The American Partisan*, May 15, 2002. http://www.american-partisan.com/cols/2002/tremlett/qtr2/0515.htm.

61 Bill Maher, "When Can We Finally Be Funny Again?" *Los Angeles Times*, September 10, 2006. http://www.latimes.com/news/printedition/opinion/la-op-maher10sep10,1,3356446.story.

62 Jim Rutenberg and Bill Carter, "Draping Newscasts with the Flag." *New York Times*, September 20, 2001.

63 Ruth Conniff, "Patriot Games," *Progressive*, January 14, 2002.

64 Conniff.

65 David Shaw, "A Skeptical Journalist Isn't an Unpatriotic One," *Los Angeles Times* (April 20, 2003): E16.

66 Albert Boime, "Waving the Red Flag and Reconstituting Old Glory," *Smithsonian Studies in American Art*, 4, 2 (1990): 2–25.

67 Elizabeth Birge and June Nicholson, "Journalism after 9/11," Quill, 92, 6 (2004): 17.

68 See, for example, http://www.newsbusters.org/blogs/scott-whitlock/2007/10/05/abc-puzzles-over-obsession-u-s-flag-pins-notes-nixon-wore-one and http://www.mrc.org/ cyberalerts/2003/cyb20030303.asp#1.

69 Cyberalert from Media Research Center, September 24, 2001 http://www.mediaresearch.org/printer/cyberalerts/2001/cyb20010924pf.asp.

70 Cyberalert from Media Research Center, October 9, 2001 http://www.mediaresearch.org/printer/cyberalerts/2001/cyb20011009pf.asp.

71 Brooke Barnett and Laura Roselle, "Patriotism in the News: Rally Round the Flag," *Electronic News*, 2 (2008): 10.

72 Barnett and Roselle, 2008.

73 Phillip Seib, *Beyond the Front Lines: How the News Media Cover a World Shaped by War* (New York: Palgrave Macmillan, 2004), 79.

74 Patrick O'Heffernan, "A Mutual Exploitation Model of Media Influence in US Foreign Policy," in W. Lance Bennett, ed. *Taken by Storm: The Media, Public Opinion and US Foreign Policy in the Gulf War* (Chicago: University of Chicago Press, 1994).

75 "Patriotism and the Press," *New York Times*, June 28, 2006 http://www.nytimes.com/2006/06/28/opinion/28Wed1.html?_r=1&oref=slogin.

76 http://query.nytimes.com/gst/fullpage.html?res=9A04E5D71430F933A05755C0A9609C8B63.

77 Richard A. Brody, *Assessing the President: The Media Elite, Elite Opinion and Public Support* (Stanford: Stanford University Press, 1991).

78 Doris Graber, "Terrorism, Censorship and the 1st Amendment: In Search of Policy Guidelines," in Montague Kern, ed. *Framing Terrorism: The News Media, the Government, and the Public* (New York: Routledge, 2003).

79 Mark Jurkowirz, "Rather Unbowed: The Departing Anchor Stands Firm on His Record and the Media's Role," *Boston Globe*, March 8, 2005.

80 Andrew Buncombe, "US Cartoonists under Pressure to Follow the Patriotic Line," *Independent*, June 23, 2002. http://www.commondreams.org/headlines02/0623-02.htm.

81 Tom Shales, "A Media Role in Selling the War? No Question." *Washington Post*, April 25, 2007.

82 Steve Schmidt, "Rumsfeld Response Due on Embedded Reporters; CNN Sniper Video Criticized, Praised," *San Diego Union-Tribune*, October 28, 2006.

83 Rutenberg and Carter.

84 Quill (April, 1996) cited in Mick Hume, *Whose War Is It Anyway? The Dangers of the Journalism of Attachment* (London: LM Special, Informc, 1996), 6.

85 William Rivers, Wilbur Schramm, and Clifford Christians, *Responsibility in Mass Communication* (New York: Harper and Row, 1980).

86 Michele Hilmes, "British Quality, American Chaos," *Radio Journal: International Studies in Broadcast and Audio Media*, 1 (2003): 13–27.

CHAPTER 8
Lessons Learned

Covering terrorism is a difficult balancing act. To be sure, the most fundamental function of the free press is its responsibility to fully inform the public. Thus, terrorism must be reported; the question is, how and how much to report. Is it in the public interest to play every act of terrorism as the day's leading news?[1]

—Political Scientist Brigitte Nacos

What is the role of a free and independent press in a democratic society? Is it to be a passive conduit responsible only for the delivery of information between a government and its people? Is it to aggressively print allegations and rumor independent of accuracy or fairness? ... No. The role of a free press is to be the people's eyes and ears, providing not just information but access, insight and, most importantly, context.

America (The Book) by Jon Stewart

Since September 11, 2001, stories about terrorism have appeared with growing frequency in the U.S. news media. Some of these stories cover terrorist events, some report on potential threats, and many try to unearth the causes of terrorism and how government seeks to combat it. As this book has shown, news coverage of terrorism is complex. Scholars from a variety of academic disciplines and journalists from across the globe have not only enriched our understanding of the challenges inherent in reporting about terrorism, but they have also highlighted the need to continue to explore the relationship between the news media and terrorism; they have also underscored the original research conducted in the context of mass media theory and with a journalism focus. As much of the contemporary terrorism research shows, one of the central questions scholars and journalists need to tackle is how to balance the requirements of a free and responsible press with the national security interests that typically arise during a terrorist attack.

This balance is harder to achieve than ever before. Even though some of the techniques used by today's terrorists date back to the first century, modern-day terrorism has evolved into a phenomenon that transcends history. Contemporary terrorists understand the power of news media and the mass audience it can cater to, so much so that some of the more organized groups have created their own media arms. As-Sahab is al Qaeda's media production house—it produces al Qaeda's video statements in various languages and often enhances the videos by altering the backgrounds using a chroma-key and digital technology.[2] This kind of "PR" heightens the concerns of scholars like Brigitte Nacos who argue that terrorists already have "a significant advantage" in getting their messages heard by the media "because their violent deeds are a powerful message that commands the mass media's attention."[3]

Meanwhile, governments are becoming increasingly media savvy themselves, a trend that started early in the twentieth century but has been refined significantly in the twenty-first century. In April 2008, an Associated Press investigation of U.S. Department of Homeland Security, the National Counter Terrorism Center, and State Department documents related to terrorism showed that these federal agencies had been circulating memos and reports to help government officials understand the appropriate language to use when talking publicly about terrorism. In a memo titled "Words that Work and Words that Don't: A Guide for Counterterrorism Communication," the Extremist Messaging Branch at the National Counter Terrorism Center emphasized that "language is critical in the war on terrorism."[4]   In a Homeland Security report, the government acknowledged that U.S. officials "may be unintentionally portraying terrorists, who lack moral and religious legitimacy, as brave fighters, legitimate soldiers or spokesmen for ordinary Muslims." The report then says, "Regarding 'jihad,' even if it is accurate to reference the term, it may not be strategic because it glamorizes terrorism, imbues terrorists with religious authority they do not have."[5]

The press is constantly navigating its relationships between the government and the audience, with both sides often claiming journalistic bias. Director of Electronic Journalism at the American University in Cairo Lawrence Pintak describes the way foreign audiences and journalists see U.S. news media:

> Journalistic bias? Like terrorism, it's in the eye of the beholder. After five years of Sturm und Drang from the Bush administration about the evils of the Arab media, American officials still don't really get it. The genie is out of the lamp. News people abroad—whether Arabs, Irish, or Zimbabweans—do see the world, and U.S. policy, differently than their American counterparts. Their news organizations will report differently. It's a fact.[6]

An ethnography of multiethnic news consumers across the United Kingdom showed distrust of Western mainstream news media. This study found that among those who turn to many news sources in more than one language, "mainstream western news is seen to be marred by ethnocentrism, to operate from within a western ideological realm and to reproduce the discursive logic of the government because journalist's access to information is seen to be dictated to a large extent by the government."[7]

The press must also deal with accusations that media cause unnecessary panic that in turn aids terrorists, something that many journalists cite as a concern. In November 2001, *Washington Post* columnist Robert Samuelson acknowledged his fear that "our new obsession with terrorism will make us its unwitting accomplices," causing journalists to become "merchants of fear." Samuelson writes,

> The perverse result is that we may become the terrorists' silent allies. Terrorism is not just about death and destruction. It's also about creating fear, sowing suspicion, undermining confidence in public leadership, provoking people—and governments—into doing things that they might not otherwise do. It is an assault as much on our psychology as on our bodies.[8]

Jamie Dettmer argues that the press is fed "scare stories" by the government, and that these stories do not serve the public interest.

> Nowadays, with 24 hour news cycles, cable and satellite TV and editorial standards that allow too much speculation and ill-informed analysis to pass as news, the pernicious side of the media's role in confrontations with terrorism has increased. With its voracious appetite needing to be satisfied, the TV media remain in hyperactive overdrive, giving the impression that the United States is on the brink of turning into a Belfast or a Beirut at the height of their troubles. The public is being scared witless.[9]

Nacos has called the three-way reciprocal communication process between the media, the public, and the government—the "Triangle of Political Communication."[10] As noted in chapters 2 and 3, Nacos argues that terrorists gain entry into the triangle through the mass media, and suggests that the government competes "with the perpetrators of political violence in that each side wants to have the loudest and most persuasive voice and message."[11] Because this book takes a press-centered approach to the study of terrorism, exploring the intersections of media/government and media/public in Nacos' triangle offers useful context that may help move the study of the relationship between the news media and terrorism forward in both academic and pragmatic ways.

Media/Government

In September 2007, Robert Adams, a citizen of Milton, Massachusetts, wrote a letter to the editor of the *New York Times*:

The White House announces that the sky is falling. The media then report either or both of two possibilities: "The sky is falling," or "The White House says the sky is falling." What one rarely reads is, "The White House says the sky is falling but it presents no real evidence to support that claim." More rarely does one see: "The White House claims that the sky is falling. Here's the evidence that suggests that's a false claim."[12]

Adam's letter suggests that what some audiences expect from the press is evidence and analysis, not just stenography—the press should make sense of what the government says and investigate official statements. Nacos argues that a softer stance is perhaps appropriate, but that softening the stance can be taken too far:

In taking a softer stand vis-à-vis the president, administration officials, members of Congress, and officials at lower level governments, the news media made the right choice when encountering a crisis that presented the country with problems it had never faced before. But suspending the adversarial stance of normal times is one thing, to join the ranks of cheerleaders is another.[13]

Others have suggested that a strong line of questioning to the government is needed specifically during a crisis and have asserted that critical analysis is both patriotic and good journalism. In testimony before a select committee on Homeland Security in the U.S. House of Representatives, former journalist and media scholar Marvin Kalb said that after 9/11 a new kind of patriotism emerged that encourages vigilant journalism:

Perhaps the new patriotism can be merged with the old kind of patriotism and that is for journalists to hold government to the old standards of truth telling, to hold announcements and proclamations up to the sunlight for confirmation of their inherent truth. For only the truth in the long run, even in this age of terrorism can really keep us free.[14]

This watchdog role is critical during crisis. Although Nacos argues that the softer stand is the right choice, we would argue that it is simply the palatable choice in light of government and audience pressure. Rigorous questioning of government officials is key, even in the aftermath of crisis.

News media's tough questioning and fulfilling the watchdog function is only one important part of the government-media relationship. Journalists must first have access to sources in order to be able to question them. From 2001 through 2005, the Reporter's Committee for Freedom of the Press

(RCFP) published a comprehensive document that detailed government actions to limit information after September 11. In *Homefront Confidential*, the RCFP notes that

> Federal Act [FOIA] officers now act under directions to give strong consideration to exemptions before handling out information, and to protect "sensitive but unclassified" information. Federal web sites have come down ... and federal courts have ruled that the government is owed deference in its [Freedom of Information] Act denials when it claims national security might be affected–even if the records are not classified.[15]

It's not just journalists who complain about the rollback in federal and state openness since September 11. Some government officials have called for the federal government to redefine freedom of information policies that would balance national security with access. At a FOIA oversight hearing in 2005, Pennsylvania representative Todd Platts, who chairs the House Subcommittee on Government Management, Finance and Accountability, criticized the large backlogs of information requests and the lack of federal agencies' enforcement of FOIA.[16]

At the end of 2007, President Bush signed the OPEN Government Act of 2007 that amended several procedural provisions of FOIA that directly address the concerns of people like Platts, but the law did not broaden the kinds of information subject to disclosure, and did not attempt to find a meaningful balance between national security and access.[17]

While journalists would argue for enhanced access to information and sources when covering terrorism in both breaking news and longer-term scenarios, some terrorism experts suggest that circumventing the news media is the best way for crisis managers to get their message directly to the people. The argument in favor of leaving the media out stems from the concern that the news media often behave irresponsibly during a crisis.[18] Instead of trusting journalists to make good choices, Nacos instead looks to crisis managers to do what they can to take the power and opportunity away from journalists. This suggestion is problematic considering the dangers of an unchecked government. Journalists are rightfully not keen on developing too close relationships with the government because of the perception that it compromises objectivity. Kalb notes, "Even if reporters wanted to believe the government, wanted to cooperate in some areas, especially now in an age of terrorism, many of them feel they cannot. They feel they must remain skeptical—in their own professional interests, but also, they feel, in the longer-range interests of the

American people."[19]

In addition to remaining appropriately "skeptical," as Kalb notes, journalists have also demonstrated the ability to use good judgment in times of crisis or when national security interests are at stake. As noted in chapter 4, most media outlets chose not to show video or still images of people jumping to their deaths from the burning World Trade Center towers immediately after the attacks. These images were not only graphic and disturbing, but also had the potential to heighten fear. Journalists publicly debated the appropriateness of showing these images and by doing so helped create greater public understanding into the journalistic decision-making process. In terms of sensitivity to national security interests, several journalism organizations held or did not run stories that could have potentially jeopardized the safety of American troops fighting the "war on terrorism" in Iraq, including the initial reporting of the Abu Ghraib prison abuse scandal.[20] That scandal in particular provides a good illustration of why many journalists are skeptical about the government, and how they weigh the public interest against some degree of respect for and deference to the government in national security situations prior to publishing or airing an important story. As noted in chapter 3, *Harper Magazine* President and Publisher John MacArthur's strong disapproval of Dan Rather's decision to withhold breaking the Abu Ghraib story gets to the heart of this kind of debate:

> [Dan Rather] was just a cheerleader for power, a cheerleader for the administration. Even now, when he goes to redeem himself with the pictures of Abu Ghraib prison, he sits on the story for two weeks. Nobody's called him on this. Because the chairman of the Joint Chiefs of Staff asked him to sit on the story? And then he says, essentially, "We only broke the story because we had to. We didn't break the story because we should, or because it is the right thing to do for the American people or for the world. We broke the story because somebody else was going to break it. It was going to come out on the Internet." This is just utter cowardice on his part.[21]

MacArthur's allegiance in this case is clearly to the public interest, but Rather was showing some sensitivity to the government's position that by breaking this story U.S. troops might be harmed in retaliation. The Abu Ghraib story came to light by accident, and even if a crisis manager or another public information officer had intentionally bypassed the media it likely would not have had any effect. Further, to suggest that journalists cannot be trusted to have important discussions that involve considering both the public interest and government well-being, but that crisis managers can, is short-sighted. To leave decisions that have tremendous impact on the American public only to crisis managers who may not have at their disposal a full spectrum of information and perspectives undermines the purpose of having a free press.

Striking a balance between the unfettered news media with absolute freedom and no concern for the effect of their messages, and state censorship and press control would be ideal for both free speech and for preventing or managing terrorism. Free speech and press is not a right in a vacuum. It is balanced with other rights such as security, both historically as well as with contemporary law and society. California Representative Christopher Cox described the relationship of the press and terrorism in a recent homeland security hearing:

> Realizing that the media plays [sic] an important role in combating terrorism does not and should not ever give license to government to control the information they provide. That said, the independence of the media should never be used as an excuse to avoid responsibility. In this spirit, the media and government can and must work constructively without necessarily working collaboratively, effectively providing uncompromised information to best serve the public.[22]

Cox suggests that media self-discipline might be the answer, although Nacos does not think that the press can be trusted to self-regulate, despite some notable examples of press restraint such as withholding rape victim names[23] and not publishing information when the government suggests it may pose a potential security threat.[24]

Further, in the wake of a terrorist attack, the public is quite willing to go the way of state-sponsored censorship, which in turns creates an economic incentive to use restraint when reporting. As Supreme Court justices have noted, there is a reason that free speech issues are not subject to vote because of the concern for the rights of others or the free press plummets during times of crisis. A CNN/Time poll found that 72 percent of Americans were fine with the government withholding information from the media or the public and 68 percent thought that the media provided too much information about the U.S. military actions.[25] This can be contrasted with the reactions of journalists like Carol Morello at the *Washington Post* who noted that during coverage of the U.S. military operations in Afghanistan that "any time news came within an inch of breaking out, we were told [by the U.S. military] that we could not report it."[26]

Media/Public

Self-imposed media guidelines on reporting about terrorism might produce what the press will consider better journalism, but the question remains if the public will see it as such.

> The reality of the new digital world means that Americans may not like what they see. These channels will show the often yawning gap between words and deeds. "We are not there to be diplomatically correct," Al Jazeera's managing director, Wadah Khanfar, recently told me. "We are there to practice journalism."[27]

Although media critics complain that journalists are falling down on the job, journalists must still contend with a public that may think that they have gone too far and reported too much. Bill Kovach argues, "In order to help the public better understand the independent role of journalists in our society and its value to them as individuals and as members of a self-governing community, journalists must create a new relationship with the public, bringing them into the processes of newsgathering."[28] Media transparency and more understanding of the significant role a critical press plays in society are two key areas to help the public understand the role the press plays in crisis.

Media transparency is all the buzz in newsrooms large and small. From full-time public reporter or ombudsman positions to simply printing e-mail addresses for readers to contact reporters, news media are finding news ways to let the public see the news process and comment on it.

At the 2005 Aspen Institute Conference on Journalism and Society, participants spent nearly half the time of the conference

> debating the merits of various measures to increase transparency, heighten accountability, and strengthen stakeholder trust in journalism ... A culture of transparency in news organizations, participants felt, will create a heightened sense of accountability by the media to their various audiences: sources and subjects of news reports, the public, employees, peers, advertisers, and shareholders.[29]

Meanwhile, others argue that the transparency efforts have gone too far and the public does not need or want to wade through every decision made in covering terrorism.[30] Participants in the Aspen Institute Conference on Journalism and Society ended their report with a note on the limits of transparency stating, "too much navel-gazing and looking backward might very well divert an organization from what it should be doing most: concentrating on explaining the world."[31] The fear that accountability efforts will hamper journalistic endeavors is almost always addressed along with the merits of transparency:

> The effort can be painful. As Kathleen Carroll, AP senior vice president, has said, "I know editors who are practically paralyzed in their newsrooms because they have been exhorted to be more responsive to their communities, but there's a community of people who just want to scream at you. They don't want to engage in a dialogue." Many a reporter whose e-mail address is published with his or her news stories says the choice becomes responding to everyone or doing the next reporting assignment. In this way, accountability and transparency, unless handled well, can be the enemy of good journalism.[32]

Former *Washington Post* ombudsman Geneva Overholser says that transparency is a relatively new phenomenon, taking place in a time that journalistic information resides indefinitely on the Web. A public reporter's discussions of media missteps can, according to *American Journalism Review* writer Rachel Smolkin, "create the misconception that the sins under discussion are new and that the media are becoming more ethically challenged by the minute."[33] This could be one factor driving the low credibility ratings for the press.

Perhaps a better solution is to get citizens directly involved in producing journalism and also perhaps enculturating the public into recognizing the need for unencumbered and rigorous terrorism coverage. In the mid-1990s, media scholars and journalists began a conversation that converged into a movement called public journalism or civic journalism. Collectively, they decided that "journalism ought to make it as easy as possible for citizens to make intelligent decisions about public affairs, and to get them carried out."[34] New York University professor Jay Rosen, considered one of the intellectual fathers of the movement, suggested that public journalism would be critically important to democracy:

> Democracy not only protects a free press, it demands a public-minded press. What democracy also demands is an active, engaged citizenry, willing to join in public debate and participate in civic affairs. No democracy—and thus, no journalist—can afford to be indifferent to trends in public (or private) life that either draw citizens toward the public sphere or repel them from it. Part of journalism's purpose, then, is to encourage civic participation, improve public debate, and enhance public life, without, of course, sacrificing the independence that a free press demands and deserves. Taken together, these propositions amount to a revised public philosophy for daily journalism.[35]

Once public journalism as a philosophy was articulated, it began slowly working its way into newsrooms across the country. By the end of the 1990s, dozens of newspapers and television stations regularly practiced public journalism, mostly in the form of giving citizens direct input into election coverage, holding town meetings, and convening focus groups to help direct coverage of important issues.[36] The news organizations that decried public journalism argued that it encouraged bias and directly challenged the concept of objectivity.[37] Public journalism proponents dismissed this claim and instead argued that the only bias public journalism encouraged was one toward democracy and public discourse.

Despite its growing success a decade ago, today public journalism as a trend has largely subsided, and has been replaced with what many call citizen journalism. Citizen journalism is practiced today by most major news organizations. By definition, citizen journalism encourages people to submit photos, videos, and other newsworthy materials to the media for publication. Citizen journalism has spread in popularity on the Internet, particularly with the creation of sites like YouTube where users can directly upload and share content. Citizen journalism also has a significant presence in mainstream news now. This became quite visible during the coverage of the London subway bombings.

One of the most published images from the London attacks was a blurry and poorly lit picture taken with a cell phone by a man who was a subway passenger trapped by the blasts. Within hours of the bombing, his picture was aired on television, published on the Web and subsequently even became a front-page newspaper photo.[38] Immediately following the London attacks, CNN, the BBC, London's The Sun, and the World Picture Network all asked bystanders to send pictures or video of the bombing.[39]

Citizen journalism also played an important role in the coverage of the Benazir Bhutto assassination in December 2007. As BBC editor Peter Horrocks discussed in his blog two weeks after Bhutto's death, "text messages and e-mails from our audiences have brought a valuable additional aspect to our journalism. But, how much attention should we pay to people who care strongly enough about an issue to send a message? They might either be typical of a wide part of the audience or perhaps just a tiny vocal minority."[40]

The rise of citizen journalism has brought discussions similar to those that surrounded public journalism as it was gaining legitimacy—does this kind of reporting enhance democracy and offer better models of news coverage that might bring journalists and the public closer to a common ground of understanding? Or, is it simply bypassing media norms and routines and encouraging new forms of bias? Max McCombs and Amy Reynolds have suggested that the purpose of public journalism was to encourage the news media to "facilitate intelligent decisions by citizens about public affairs and that they should involve citizens in the process of deciding what issues are most important for both the media and the public agenda."[41] As this book has documented, since September 11, terrorism and the resulting war against it has been a top priority of the news media, government and the public, which might create an opportunity for all three players in Nacos' political communication triangle to find common ground.

Conclusion

As noted in the introduction, one of the goals of this book was to present research about the relationship between terrorism and the press through a news media lens that might enhance interdisciplinary understanding. Previous research tells us that the media is the primary vehicle through which a significant number of people learn about and come to understand terrorism. This should lead mass communication scholars to consider how the press reports on terrorism, how that reporting varies depending on the medium and globally and within different media and government systems. These are some of the issues this book has addressed, through the application of a variety of mass communication theories and methods.

In addition to using a press-centered approach to explore the news media and its relationship to terrorism, this book has also sought to highlight terrorism-related research within the field of journalism and mass communication; expand on the terrorism-related research from other fields to incorporate the press perspective; and to provide original research that better explains how the news media view the so-called symbiotic relationship between the media and terrorism. This kind of approach to research about terrorism and the media should continue to offer value to both scholars and to journalists.

Future studies should continue to examine news media coverage of terrorism in terms of how the press, audiences, and governments interact and how their different roles inform and impact coverage. The three corners of Nacos' triangle deserve more research, specifically focused on the ways the three interact and how context shapes their interactions. One theory not addressed in this book that could offer additional insights into the media and terrorism relationship is performance theory. Performance theory applied to terrorism news could explore the rich intersection between differing expectations of the news report by the news media, the government, and the public and how the convergence of performer, event, setting, and audience provide important contexts for understanding journalistic "texts." Richard Bauman helped define the concept of performance theory, arguing that oral literature analysis needed to move beyond the analysis of the texts:

> The texts we are accustomed to viewing as the raw materials of oral literature are merely the thin and partial record of deeply situated human behavior. My concern has been to go beyond a conception of oral literature as disembodied superorganic stuff and to view it contextually and ethnographically, in order to discover the individual, social, and cultural factors that give it shape and meaning in the conduct of social life.

Using performance theory could enhance our understanding of how terrorism news coverage is dependent on the interrelationships among journalists, government, and audience, thus ensuring proper examination of how interviews and news stories are framed, including discourse and interaction before and after statements that are content analyzed. Better understanding of how terrorism is covered across cultures is also worthy of exploration. And finally, a clearer sense is needed of how new media are changing the way terrorism is covered in the press, how governments get their message out after a terrorism attack, and how the public receives both of those messages.

The exploration of the relationship between the media and terrorism is so complex that our ultimate understanding, if possible, will require continued multidisciplinary research and the inclusion of diverse perspectives. It may also require adjustments away from traditional thinking, practice, and theory as terrorists, governments, and the media continue to adapt to the changing global society, new advances in technology, and both predictable and unpredictable events that will unfold as the twenty-first century moves forward.

A close reading of the research, theory, and practice discussed in this book leads to three final recommendations. The first underscores the need for the owners of major media companies to consider the price that society may pay as media companies engage in continued cost-cutting. As these pages head off to the typesetter, yet another major media company is announcing job cuts. McClatchy Company, the country's third-largest newspaper publisher, announced on June 17, 2008, that it would lay off 10 percent of its workforce.[42] This announcement reflects a trend in newspapers downsizing. In 2005 the

New York Times cut its work force by five hundred employees, including eighty newsroom positions and the Knight Ridder Company cut hundred newsroom employees in Philadelphia.[43]

Journalism's business side is increasingly engaging in cost-cutting, job-eliminating, and bureau-closing. This could significantly damage the "product" and result in lower quality journalism. Fewer journalists will cover fewer stories and will typically have less time to devote to doing it. For terrorism coverage, this will likely mean even less specialized or nuanced coverage from reporters in the field who understand the context in which terrorism occurs. The recent lead up to the war in Iraq is an example of how journalism can fall down on the job. With the dissolving of foreign bureaus, news organizations have fewer foreign correspondents. This means that vast regions of the world are often unrepresented in U.S. news. Often, in these areas of the world where foreign bureaus do not exist or have been drastically cut, the focus of news coverage comes only when a crisis hits. This means that there are no established journalists with contacts and understanding in the crisis region who can start deep reporting immediately. Instead, reporters are flown into a region to figure things out, while simultaneously trying to report on them. It is the equivalent to a tourist view of conflict. Media owners need to understand that journalism cannot be a business only concerned with the bottom line. The quality of coverage matters, especially because this is often the only view of the world and of terrorism that many people will ever see.

The second recommendation is for individual journalists who are increasingly put in a difficult position to cover terrorism effectively. Reporting "beats" have increasingly been combined, and as a result coverage of some issues has diminished. In an increasingly market-driven environment, what people want in terms of entertainment will often trump what they need in terms of democracy. Journalists must fight the tendency to think about the bottom line when reporting on terrorism. A staggeringly high number of journalists already feel that bottom line pressure affects the quality of journalism. In a 2004 survey, 66 percent of national news people and 57 percent of local journalists said that increased bottom line pressure is "seriously hurting" the quality of news coverage.[44] Collective bargaining around this issue is notably hard during a climate of cut backs, but this is precisely what journalists must strive to do. Journalists must continue fighting for the fire wall between editorial and advertising, between journalists and business managers. The government is getting more savvy about media as the media are employing more citizen voices at the expense of hiring more experts. The impact of this can be felt by increasingly distracted audiences who avoid critical analysis of the press or the information that the press presents. The war in Iraq was a failure of both journalists and audiences critically examining government and the information about terrorism available at the time. This is a tremendous responsibility that should not be underestimated and cannot be managed by people only concerned with financial bottom lines.

The last recommendation concerns the role of the audience, who must back the media owners and journalists who do try to bring serious and crucial news about terrorism into public consciousness. Journalists who critically focus on issues and ask the important and difficult questions must not be called unpatriotic and should not have their reporting dismissed out-of-hand. During the lead up to the Iraq War, journalists should have been more savvy about questioning the government's assertion about the connection of the terrorism attacks of September 11 to Saddam Hussein. Journalists should have asked harder questions; they should have pressed the government officials who were presenting the case for war. And those who own the media outlets should have been brave enough to provide resources and support for those stories. The audience should have been mature enough to handle hearing, watching, and reading these stories. Better terrorism coverage requires everyone to raise their expectations of each other—the press, the government, and the audience—and to continually hold each other accountable.

Notes

1   Brigitte Nacos, *Mass-Mediated Terrorism: The Central Role of the Media in Terrorism and Counterterrorism* (2nd ed.) (Lanham, Md.: Rowman & Littlefield, 2007), 227.
2   Neal Krawetz, "A Picture's Worth ... Digital Image Analysis and Forensics." Paper presented at Black Hat Briefings, 2007.
3   Nacos, 197.
4   Matthew Lee, "Government Changing Terrorism Language," available online at http://www.time.com/time/nation/article/0,8599,1734909,00.html?xid=feed-cnn-topics [Accessed April 28, 2008].W
5   Lee.
6   Lawrence Pintak, "America's Media Bubble: A Willful Blindness," *International Herald Tribune*, November 19, 2006 http://www.iht.com/articles/2006/11/19/opinion/edpintak.php
7   Marie Gillespie, "Shifting Securities: News Cultures before and beyond the Iraq War 2003," http://www.mediatingsecurity.com/
8   Robert J. Samuelson, "Unwitting Accomplices?" *The Washington Post*, November 7, 2001, A29.

9   Jamie Dettmer, "Supplying Terrorists the 'Oxygen of Publicity,'" *Insight*, July 15, 2002, 47.

10 Nacos, 15.

11 Nacos, 197.

12 Found at this Web site http://query.nytimes.com/gst/fullpage.html?res
    =9502EFD9103EF933A0575AC0A9619C8B63&sec=&spon=&pagewanted=2

13 Brigitte Nacos, "Terrorism as Breaking News: Attack on America," *Political Science Quarterly*
    (Spring 2003): 23–52.

14 Marvin Kalb, "Testimony before the Select Committee of Homeland Security in the U.S.
    House of Representatives," September 15, 2004, 7

15 Reporters Committee for Freedom of the Press, *Homefront Confidential: How the War on Terrorism
    Affects Access to Information and the Public's Right to Know* (6th ed.), available online at
    http://www.rcfp.org/homefrontconfidential/ [Accessed May 2, 2008].

16 Reporters Committee for Freedom of the Press.

17 Douglas Lee, "What's on the Horizon," available online at
    http://www.firstamendmentcenter.org/press/information/horizon.aspx?topic=FOI_horiz
    on [Accessed May 1, 2008].

18 Nacos, *Mass-Mediated Terrorism* ,212.

19 Kalb, 7

20 John MacArthur, "Everybody Wants to Be at Versailles," in Kristina Borjesson, ed. *Feet to the
    Fire: The Media after 9/11* (Amherst, N.Y.: Prometheus Books, 2005), 92–122.

21 MacArthur, 105.

22 Christopher Cox, "Opening Statement before the Select Committee of Homeland Security in
    the U.S. House of Representatives," September 15, 2004, 3.

23 Christy Oglesby, "Rape Victims' Names Withheld by Choice, Not Law: Statutes on Confi-
    dentiality Don't Trump Media's Constitutional Rights," CNN Web site October 16, 2003,
    http://www.cnn.com/2003/LAW/10/16/rape.confidential/

24 Dean Baquet and Bill Keller, "When Do We Publish a Secret?" *Los Angeles Times* and *New York
    Times*, July 1, 2006.

25 Media Research Cyber Alert from November 19, 2001   www.mediaresearch.org/_cyber-
    alerts/2001/cyb20010019.asp

26 Stephen Hess and Marvin Kalb, eds., *The Media and the War on Terrorism*, (Washington, D.C.:
    Brookings Institution Press, 2003), 167.

27 Lawrence Pintak, "America's Media Bubble: A Willful Blindness," *International Herald Tribune*,
    November 19, 2006 http://www.iht.com/articles/2006/11/19/opinion/edpintak.php

28 Bill Kovach, "Journalism and Patriotism," *Sala de Prensa*, 46, 2 (2002),
    http://www.saladeprensa.org/art381.htm

29 Jon Ziomek , *Journalism, Transparency and the Public Trust*: A Report of the Eighth Annual Aspen
    Institute Conference on Journalism and Society, 2005. www.aspeninstitute.org/
    atf/cf/%7BDEB6F227-659B-4EC8-8F84-8DF23CA704F5%7D/JOURTRANSPTEXT.PDF -

30 Rachel Smolkin, "Too Transparent?" *American Journalism Review* (April/May 2006),
    http://www.ajr.org/Article.asp?id=4073

31 Ziomek.

32 Geneva Overholser, "On Behalf of Journalism: A Manifesto for Change,"
    www.appcpenn.org/Overholser/20061011_JournStudy.pdf

33 Smolkin,

34 Arthur Charity, *Doing Public Journalism* (New York: Guilford, 1995), 2.

35 Jay Rosen, "Community Connectedness: Passwords for Public Journalism," *Poynter Report*, 3 (1993): 3–16, 3.

36 Amy Reynolds, "Local Television Coverage of the NIC," in Maxwell E. McCombs and Amy Reynolds, eds. *The Poll with a Human Face: The National Issues Convention Experiment in Political Communication* (Mahwah, N.J.: Lawrence Erlbaum Associates, 1999).

37 James Fallows, *Breaking the News: How the Media Undermine American Democracy* (New York: Pantheon, 1996).

38 Yuki Noguchi, "Camera Phones Lend Immediacy to Images of Disaster," available online at http://www.washingtonpost.com/wp-dyn/content/article/2005/07/07/AR2005070701 522.html [Accessed May 2, 2008].

39 Noguchi.

40 Peter Horrocks, "Value of Citizen Journalism," available online at http://www.bbc.co.uk/blogs/theeditors/2008/01/value_of_citizen_journalism.html [Accessed May 2, 2008].

41 Maxwell E. McCombs and Amy Reynolds, "Enhancing Grassroots Democracy," in Maxwell E. McCombs and Amy Reynolds, eds. *The Poll with a Human Face: The National Issues Convention Experiment in Political Communication* (Mahwah, N.J.: Lawrence Erlbaum Associates, 1999), 211–220, 215.

42 Mark Fitzgerald, "Behind McClatchy Layoffs—A Mountain of Debt," *Editor and Publisher* online edition, June 17, 2008. http://www.editorandpublisher.com/eandp/news/article_display.jsp?vnu_content_id=1003817340

43 Katharine Q. Seelye, "Times Company Announces 500 Job Cuts," *New York Times,* September 21, 2005, http://www.nytimes.com/2005/09/21/business/media/21paper.html?_r=1&oref=slogin

44 The Pew Research Center for the People and the Press, "Bottom-Line Pressures Now Hurting Coverage, Say Journalists Press Going Too Easy on Bush," May 23, 2004, http://people-press.org/report/214/

BIBLIOGRAPHY

Adams, Paul, "American Interest in Iraq Slumps," BBC News Monday, March 24, 2008, http://news.bbc.co.uk/2/hi/middle_east/7311814.stm.

Adamson, Walter L., *Hegemony and Revolution: A Study of Antonio Gramsci's Political and Cultural Theory* (Berkeley: University of California Press, 1980).

Aday, Sean, Steven Livingston, and Maeve Hebert, "Embedding the Truth: A Cross-cultural Analysis of Objectivity and Television Coverage of the Iraq War," *Harvard International Journal Of Press/Politics*, 10, 1 (2005): 3–21.

Alexander, Yonah and Robert Picard, *In the Camera's Eye: News Coverage of Terrorist Events* (Washington, D.C.: Brassey's, 1991).

Altheide, David L., *Creating Fear: News and the Construction of Crisis* (New York: Aldine de Gruyter, 2002).

———, *Creating Reality: How TV News Distorts Events* (London: Sage, 1976).

———, "Media Hegemony: A Failure of Perspective," *Public Opinion Quarterly*, 48 (1984): 476–490.

———, *Terrorism and the Politics of Fear* (Lanham, Md.: AltaMira Press, 2006)

Altheide, David L. and Robert P. Snow, *Mediaworlds Postjournalism* (Berlin: Mounton de Gruyter, 1991).

American Society of Newspaper Editors, "ASNE Statement of Principles," available online at http://www.asne.org/index.cfm?ID=888 [Accessed May 1, 2008].

Anderson, Perry, "The Antinomies of Antonio Gramsci," *New Left Review*, 100 (1976): 5–78.

Andrews, Wayne, ed., *The Autobiography of Theodore Roosevelt, Condensed from the Original Edition, Supplemented by Letters, Speeches, and Other Writings* (New York: Charles Scribner's Sons, 1913, rep. 1958).

Anker, Elisabeth, "Villains, Victims and Heroes: Melodrama, Media, and September 11," *Journal of Communication*, 55, 1 (2005): 22–37.

Armstrong, Karen, "The Label of Catholic Terror Was Never Used about the IRA," http://www.guardian.co.uk./politics/2005/jul/11/northernireland.july7 [Accessed April 16, 2008].

Ashcroft, John, *Memorandum for Heads of all Federal Departments and Agencies*, available online at www.usdoj.gov/oip/foiapost/2001foiapost19.htm [Accessed May 1, 2008].

————, *U.S. Department of Justice News Conference with Attorney General Ashcroft*, available online at us-info.state.gov/topical/pol/terror/01112711.htm [Accessed May 1, 2008].

Baker, Russ, "Want to Be a Patriot? Do Your Job." *Columbia Journalism Review* (May/June 2002): 41.

Bamford, James, "Dangerous Nonsense," in Kristina Borjesson, ed. *Feet to the Fire: The Media after 9/11* (Amherst, N.Y.: Prometheus Books, 2005), 323.

Baquet, Dean and Bill Keller, "When Do We Publish a Secret?" *The Los Angeles Times* and *The New York Times*, July 1, 2006 http://www.nytimes.com/2006/07/01/opinion/01keller.html?_r=1&oref=login.

Barnett, Brooke, *The War on Terror and the Wars in Iraq* (Westport, Conn.: Greenwood Press, 2005).

Barnett, Brooke and Laura Roselle, "Patriotism in the News: Rally Round the Flag," *Electronic News*, 2 (2008): 10.

Barnett, Brooke and Maria Elizabeth Grabe, "The Impact of Slow Motion Video on Viewer Evaluations of Television News Stories," *Visual Communication Quarterly*, 7 (2000): 4–7.

Baum, Matthew A., *Soft News Goes to War* (Princeton, N.J.: Princeton University Press, 2003).

BBC Editorial Guidelines: War, Terror and Emergencies, available online at http://www.bbc.co.uk/guidelines/editorialguidelines/edguide/war/mandatoryreferr.shtml [Accessed March 14, 2008].

Begin, Menachem, "Freedom Fighters and Terrorists," in Benjamin Netanyahu, ed. *International Terrorism: Challenge and Response* (Edison, N.J.: Transaction Books, 1981), 39–46.

————, *The Revolt: Story of the Irgun* (New York: Henry Schuman, 1951).

Behr, Roy L. and Shanto Iyengar, "Television News, Real-World Cues, and Changes in the Public Agenda," *Public Opinion Quarterly*, 49 (1985): 38–57.

Berkowitz, Dan and Yehiel Limor, "A Cross-cultural Look at Serving the Public Interest: American and Israeli Journalists Consider Ethical Scenarios," *Journalism: Theory, Practice & Criticism*, 5, 2 (2004): 159–181.

Birge, Elizabeth and June Nicholson, "Journalism after 9/11," *Quill*, 92, 6 (2004): 17.

Bohlen, Celestine, "Think Tank; in New War on Terrorism, Words Are Weapons, Too," *New York Times*, September 21, 2001.

Boime, Albert, "Waving the Red Flag and Reconstituting Old Glory," *Smithsonian Studies in American Art*, 4, 2 (1990): 2–25.

Borden, Sandra L., "Communitarian Journalism and Flag Displays after September 11: An Ethical Critique," *Journal of Communication Inquiry*, 29, 1 (2005): 30–46.

Boswell, James, "Entry for Friday, April 7, 1775," *Life of Johnson* (New York: Everyman's Library, 1993), 615.

Branzburg v. Hayes, 408 U.S. 665 (1972).

Breed, Warren, "Social Control in the Newsroom: A Functional Analysis," *Social Forces*, 33 (1955): 326–335.

Broder, David S. and Dan Balz, "How Common Ground of 9/11 Gave Way to Partisan Split, *The Washington Post*, Jul 16, 2006, A01.

Brody, Richard A., *Assessing the President: The Media Elite, Elite Opinion and Public Support* (Stanford: Stanford University Press, 1991).

Brown, Robin, "Clausewitz in the Age of CNN: Rethinking the Military-Media Relationship," in Pippa Norris, Montague Kern, and Marion Just, eds. *Framing Terrorism: The News Media, the Government and the Public* (New York: Routledge, 2003).

Buncombe, Andrew, "US Cartoonists under Pressure to Follow the Patriotic Line," *Independent*, June 23, 2002. http://www.commondreams.org/headlines02/0623-02.htm.

Bykofsky, Stu, "Cable News Too Fast, Not Final," *Philadelphia Daily News*, October 10, 2006, from http://go.philly.com/byko.

Calvert, Peter, "Theories of Insurgency and Terrorism: Introduction," in the *International Encyclopedia of Terrorism* (Chicago: Fitzroy Dearborn, 1997), 135–144.

Capital Hill Hearing, "Preserving Freedoms while Defending against Terrorism: Panel 1 of a Hearing by the Senate Judiciary Committee," *Federal News Service*, November 28, 2001, 1–49.

Carter, Bill and Felicity Barringer, "A Nation Challenged: Speech and Expression; in Patriotic Time, Dissent Is Muted," *New York Times*, September 28, 2001.

Cate, Fred H., "'CNN Effect' Is Not Clear-Cut," *Humanitarian Affairs Review* (Summer 2002), available at http://globalpolicy.org/ngos/aid/2002/summercnn.htm.

Center for National Security Studies et al. v. Department of Justice (D.C. Dist. Court, December 5, 2001).

Center for National Security Studies et al. v. U.S. Department of Justice, 356 U.S. App. D.C. 333 (2003).

Chaliand, Gérard and Arnaud Blin, eds., *The History of Terrorism: From Antiquity to Al Qaeda* (Berkeley: University of California Press, 2007).

———, "Manifestations of Terror through the Ages," in Gérard Chaliand and Arnaud Blin, eds. *The History of Terrorism: From Antiquity to Al Qaeda* (Berkeley: University of California Press, 2007), 79-92.

Chermak, Steven, "The Presentation of Terrorism in the News" (Paper presented at the International Conference on the TV Presentation of Crime in Milan, Italy, May 15–16, 2003).

Chomsky, Noam, *The Culture of Terrorism* (Boston, Mass.: South End Press, 1988).

Chyi, Hsiang Iris and Maxwell McCombs, "Media Salience and the Process of Framing: Coverage of the Columbine School Shootings," *Journalism and Mass Communication Quarterly*, 81, 1 (2004): 22–35.

CNN, "Benazir Bhutto Assassinated," http://www.cnn.com/SPECIALS/2007/news/benazir.bhutto/index.html [Accessed January 15, 2008].

Cockburn, Alexander and Jeffrey St. Clair, *End Times: The Death of the Fourth Estate* (Petrolia, Calif.: Counter Punch, 2007).

Cohen-Almagor, Raphael, "Media Coverage of Acts of Terrorism: Troubling Episodes and Suggested Guidelines." *Canadian Journal of Communication*, 30, 3 (2005): 383–409.

Cohen, Joshua and Martha C. Nussbaum, *For Love of Country: Debating the Limits of Patriotism*, Beacon Press (Boston: Beacon Press, 1996).

Committee to Protect Journalists, "Iraq: Journalists in Danger," available online at http://www.cpj.org/Briefings/Iraq/Iraq_danger.html [Accessed April 17, 2008].

Conniff, Ruth, "Patriot Games," *Progressive*, January 2002, 14.

Cooke, Tim, "Paramilitaries and the Press in Northern Ireland," in Pippa Norris, Montague Kern, and Marion Just, eds., *Framing Terrorism: The News Media, the Government, and the Public* (New York: Routledge, 2003), 75–90.

Copeland, David A., *The Idea of a Free Press: The Enlightenment and Its Unruly Legacy* (Evanston, Ill.: Northwestern University Press, 2006).

Cox, Christopher, "Opening Statement before the Select Committee of Homeland Security in the U.S. House of Representatives," September 15, 2004.

Craft, Stephanie and Wayne Wanta, "U.S. Public Concerns in the Aftermath of 9/11: A Test of Second Level Agenda Setting," *International Journal of Public Opinion Research*, 16, 1 (2004): 456–463.

Craig, Frank, "Why Publish Images of Death," *Pittsburgh Tribune-Review*, Tuesday, June 22, 2004. http://www.pittsburghlive.com/x/pittsburghtrib/s_199849.html.

Crenshaw, Martha, ed., *Terrorism, Legitimacy, and Power: The Consequences of Political Violence* (Middletown, Conn.: Wesleyan University Press, 1983).

———, ed., *Terrorism in Context* (University Park, Pa.: Pennsylvania State University Press, 1995).

Dayan, David and Elihu Katz, *Media Events: The Live Broadcasting of History* (Cambridge, Mass.: Harvard University Press, 1992).

Detenber, Benjamin H. and Byron Reeves, "A Bio-informational Theory of Emotion: Motion and Image Size Effects on Viewers," *Journal of Communication*, 46, 3 (1996): 66–84.

Dettmer, Jamie, "Supplying Terrorists the 'Oxygen of Publicity,'" *Insight*, July 15, 2002, 47.

Dicken-Garcia, Hazel, *Journalistic Standards in Nineteenth-Century America*. (Madison, Wis.: University of Wisconsin Press, 1989).

Dimitrova, Daniela, Lynda Lee Kaid, Andrew Paul Williams, and Kaye D. Trammell, "War on the Web: The Immediate News Framing of Gulf War II," *Harvard International Journal of Press/Politics*, 10, 1 (2005): 22–44.

Dobkin, Bethami A., *Tales of Terror: Television News and the Construction of the Terrorist Threat* (New York: Praeger, 1992).

Drew, Dan G. "Roles and Decisions of Three Television Beat Reporters," *Journal of Broadcasting*, 16 (1972): 165–173.

Drew, Dan G. and Thomas Grimes, "Audio-Visual Redundancy and TV News Recall," *Communication Research*, 14 (1987): 452–461.

Dyson, Michael Eric, *Come Hell or High Water: Hurricane Katrina and the Color of Disaster* (New York: Basic Civitas Books, 2006).

El-Nawawy, Mohammed and Adel Iskandar, *Al-Jazeera: The Story of the Network That Is Rattling Governments and Redefining Modern Journalism* (Cambridge, Mass.: Westview Press, 2003).

Emerson, Thomas, *The System of Freedom of Expression* (New York: Vintage Books, 1970).

Entman, Robert M., "Framing: Toward Clarification of a Fractured Paradigm," *Journal of Communication*, 43 (1993): 51–58.

———, "Framing U.S. Coverage of International News: Contrasts in Narratives of the KAL and Iran Air Incidents," *Journal of Communication*, 4, 4 (1991): 6–27.

———, *Projections of Power: Framing News, Public Opinion and U.S. Foreign Policy* (Chicago: University of Chicago Press, 2004).

Fahmy, Shahira and Daekyung Kim, "Picturing the Iraq War: Constructing Images of Conflict Revisited" (Paper presented at the annual meeting of the International Communication Association, Seoul, Korea, July, 2002).

Fahmy, Shahira, Sooyoung Cho, Wayne Wanta, and Yonghoi Song, "Visual Agenda Setting After: Individual Emotion, Recall and Concern about Terrorism," *Visual Communication Quarterly*, 13 (2006): 7.

Fishman, Mark, *Manufacturing the News* (Austin, Tex.: University of Texas Press, 1980).

Fisk, Robert, "Remember 'the Whys,'" in David Wallis, ed. *Killed: Great Journalism Too Hot to Print* (New York: Nation Books, 2004), 377.

Frey, Bruno S. and Dominic Rohner, "Blood and Ink! The Common-Interest Game between the Terrorists and the Media," Public Choice, 133, 1–2 (October, 2007), 129-145.

Fox, Julia R., Glory Koloen, and Volkan Sahin, "No Joke: A Comparison of Substance in the Daily Show with Jon Stewart and Broadcast Network Television Coverage of the 2004 Presidential Election Campaign," Journal of Broadcasting & Electronic Media, 51, 2 (2007): 213.

Freedom of Information Act, 5 U.S.C, sect. 552 (1988).

Gadarian, Shana Kushner, "Beyond the Water's Edge: Polarized Reactions to Images in the War on Terror" (Paper presented at the 2007 annual meeting of the Midwest Political Science Association, Chicago, Ill. April 12–15, 2007).

Gamson, William A, "Foreword," in Stephen D. Reese, Oscar H. Gandy, Jr., and E. August, eds. Grant Framing Public Life: Perspectives on Media and Our Understanding of the Social World (Mahwah, N.J.: Lawrence Erlbaum Associates, 2001).

Gans, Herbert J., Deciding What's News—A Study of CBS Evening News, NBC nightly News (New York: Random House, 1979).

Gavel, Doug "CNN's Woodruff Assesses TV News: Veteran Journalist Gives Television High Grades for Crisis Coverage," Harvard University Gazette, November 8, 2001, http://www.hno.harvard.edu/gazette/2001/11.08/10-woodruff.html.

Gerbner, George, Larry Gross, Michael Morgan, and Linda Signorielli, "The 'Mainstreaming' of America: Violence Profile No. 11," Journal of Communication, 40, 2 (1980): 172–199.

Gilboa, Eytan, "The CNN Effect: The Search for a Communication Theory of International Relations," Political Communication, 22 (2005): 27-44.

Gillespie, Marie, "Shifting Securities: News Cultures Before and Beyond the Iraq War 2003," http://www.mediatingsecurity.com/.

Goldman, Emma, "Patriotism: A Menace to Liberty," in Anarchism and Other Essays (New York: Dover, 1969).

Goldstein, Norm, ed., The Associated Press Stylebook (New York: Associated Press, 2007).

Gomberg, Paul, "Patriotism Is Like Racism," in Igor Primoratz, ed. Patriotism (Amherst, N.Y.: Humanity Books, 2002), 105–112.

Grabe, Maria Elizabeth, "The South African Broadcasting Corporation's Coverage of the 1987 and 1989 Elections: The Matter of Visual Bias," Journal of Broadcasting and Electronic Media, 40 (1996): 153–179.

Graber, Doris, "Introduction: Perspectives on Presidential Linkage," in Doris Graber, ed. The President and the Public (Philadelphia: Institute for the Study of Human Issues, 1982), 1–14.

———, Mass Media and American Politics (7th ed.) (Washington, D.C.: CQ Press, 2006).

——, *Processing the News: How People Tame the Information Tide* (New York: Longman, 1988).

——, "Say It with Pictures," *Annals of the American Academy of Political and Social Sciences*, 546 (1996): 85–96.

——, "Terrorism, Censorship and the 1st Amendment: In Search of Policy Guidelines," in Montague Kern, ed. *Framing Terrorism: The News Media, the Government, and the Public* (New York: Routledge, 2003).

Greenberg, Bradley S. and Marcia Thomson, *Communication and Terrorism: Public and Media Responses to 9/11* (Cresskill, N.J., Hampton Press, 2002).

Gunter, Barrie, *Poor Reception: Misunderstanding and Forgetting Broadcast News* (Hillsdale, N.J.: Lawrence Erlbaum Associates, 1987).

Gunter, Barry, "Remembering Television News: Effects of Picture Content," *Journal of General Psychology*, 102 (1980): 127–133.

Gunter, Barry, Adrian Furnham, and Gillian Gietson, "Memory for the News as Function of the Channel of Communication. Human Learning," 3 (1984): 265–271.

Habermas, Jürgen, "Appendix II: Citizenship and National Identity," in William Rehg, trans. *Between Facts and Norms: Contributions to a Discourse Theory of Law and Democracy* (Cambridge, Mass.: MIT Press, 1996).

Hallin, Daniel, *"The Uncensored War": The Media and Vietnam* (New York: Oxford University Press, 1986).

Hart-Teeter, "From the Home Front to the Front Lines: America Speaks Out about Homeland Security" (March 2004), 1-51.

Hedges, Chris, "We're Not Mother Teresas," in Kristina Borjesson, ed. *Feet to the Fire: The Media After 9/11* (Amherst, N.Y.: Prometheus Books, 2005), 531–32.

Heldman, Caroline, "'Burnin' and 'Lootin': Race and Media Coverage of Hurricane Katrina" (Paper presented at the annual meeting of the Western Political Science Association, La Riviera Hotel, Las Vegas, Nevada, Mar 08, 2007).

Hertog, James K. and Douglas M. McLeod, "A Multiperspectival Approach to Framing Analysis: A Field Guide," in Stephen D. Reese, Oscar H. Gandy, Jr., and August E. Grant, eds. *Framing Public Life: Perspectives on Media and Our Understanding of the Social World* (Mahwah, N.J.: Lawrence Erlbaum Associates, 2001), 139–161.

Hess, Stephen and Marvin Kalb, eds., *The Media and the War on Terrorism* (Washington, D.C.: Brookings Institution Press, 2003).

Hilmes, Michele, "British Quality, American Chaos," *Radio Journal: International Studies in Broadcast and Audio Media*, 1, 1 (2003): 13–27.

Hirsch, Paul, "The "Scary World" of the Nonviewer and Other Anomalies: A Reanalysis of Gerbner et. al.'s Findings on Cultivation Analysis," *Communication Research*, 7, (1980): 403–456.

Hoffman, Bruce, *Inside Terrorism* (New York: Columbia University Press, 2006).

Honestreporting.com, "Editors Consider the T-word," available online at http://www. honestreporting.com/articles/critiques/Editors_Consider__the_-T-word-.asp [Accessed March 14, 2008].

Hubac-Occhipinti, Olivier, "Anarchist Terrorists of the Nineteenth Century," in Gérard Chaliand and Arnaud Blin, eds. *The History of Terrorism: From Antiquity to Al Qaeda* (Berkeley: University of California Press, 2007), 113–131.

Huck, Peter, "We Had 50 Images within an Hour,'" *The Guardian*, July 11, 2005.

Huddy, Leonie, Stanley Feldman, Gallya Lahav, and Charles Taber, "Fear and Terrorism: Psychological Reactions to 9/11," in Pippa Norris, Montague Kern, and Marion Just, eds. *Framing Terrorism: The News Media, the Government, and the Public* (New York: Routledge, 2003), 281–302.

Hunt, Gaillard, ed., *The Writings of James Madison* (New York: G.P. Putnam's Sons, 1910).

Iyengar, Shanto, *Is Anyone Responsible? How Television Frames Political Issues* (Chicago: University of Chicago Press, 1991).

Jasperson, Amy E. and Mansour O. El-Kikhia, "CNN and al Jazeera's Media Coverage of America's War in Afghanistan," in Pippa Norris, Montague Kern, and Marion Just, eds. *Framing Terrorism: The News Media, the Government, and the Public* (New York: Routledge, 2003), 113–132.

Jaszi, Oscar and John D. Lewis, *Against the Tyrant: The Tradition and Theory of Tyrannicide* (Glencoe, Ill.: Free Press, 1957), 3–96.

Jenkins, Brian, "International Terrorism: A New Mode of Conflict," in David Carlton and Carlo Schaerf, eds. *International Terrorism and World Security* (London: Croom Helm, 1975).

Jensen, Carl, ed., *Stories that Changed America: Muckrakers of the 20th Century* (New York: Seven Stories Press, 2000).

Jensen, Robert J., *Citizens of the Empire: The Struggle to Claim our Humanity* (San Francisco: City Lights Books, 2004).

———, "Journalism Should Never Yield to 'Patriotism,'" Published on Wednesday, May 29, 2002 in the Long Island, New York *Newsday*. Retrieved from the commondreams.org Web site http://www.commondreams.org/views02/0529-02.htm.

Johnson, Samuel, *Idler* #30 (November 11, 1758) as quoted on http://www. samueljohnson.com/patrioti.html.

Johnstone, John W.C., Edward J. Slawski, and William W. Bowman, *The News People: A Sociological Portrait of American Journalists and Their Work* (Urbana, Ill.: University of Illinois Press, 1976).

————, "The Professional Values of American Newsmen," *Public Opinion Quarterly*, 36 (winter 1972–1973): 522–540.

Jones, Colin, "Terror in the French Revolution 1789-1815," in the *International Encyclopedia of Terrorism* (Chicago: Fitzroy Dearborn, 1997), 48–51.

Jurkowirz, Mark, "Rather Unbowed: The Departing Anchor Stands Firm on His Record and the Media's Role," *Boston Globe*, March 8, 2005.

Kalb, Marvin, "Testimony before the Select Committee of Homeland Security in the U.S. House of Representatives," September 15, 2004.

Kennedy, Dan. "The Daniel Pearl Video: A Journalist Explains Why Its Horrific Images Should Be Treated as News," *Nieman Reports* (Fall 2002): 80.

Kern, Montague, "Introduction," in W. L. Bennett and D. Paletz D., eds. *Taken by Storm: The Media, Public Opinion and US Foreign Policy in the Gulf War* (Chicago: University of Chicago Press, 1994).

Kern, Montague, Marion Just, and Pippa Norris, "The Lessons of Framing Terrorism," in Pippa Norris, Montague Kern, and Marion Just, eds. *Framing Terrorism: The News Media, the Government, and the Public* (New York: Routledge, 2003), 281–302.

Kim, Yung-Soo and Zoe Smith, "News Images of the Terrorist Attacks: Framing September 11th and Its Aftermath through the Pictures of the Year International Competition" (Paper presented at the annual meeting of the AEJMC, Kansas City, Mo., August 2003).

Kiousis, Spiro, Michael Mitrook, Trenton Seltzer, Cristina Popescu, and Arlana Shields, "First- and Second-Level Agenda Building and Agenda Setting: Terrorism, the President and the Media" (Paper presented at the International Communication Association, June 16, 2006, Dresden, Germany).

Kobre, Sidney, *The Yellow Press and Gilded Age Journalism* (Tampa, Fla.: Florida State University, 1964).

Kovach, Bill, "Journalism and Patriotism," *Sala de Prensa*, 46 (2002): 2.

Krawetz, Neal, "A Picture's Worth … Digital Image Analysis and Forensics" (Paper presented at Black Hat Briefings, 2007).

Laqueur, Walter, *The Age of Terrorism* (Boston: Little, Brown, 1987).

Lasswell, Harold, *Propaganda Technique in the World War* (New York: Peter Smith, 1927).

Lazarsfeld, Paul, Bernard Berelson, and Hazel Gaudet, *The People's Choice* (New York: Columbia University Press, 1944).

Lee, Alfred McClung and Elizabeth Briant Lee, eds., *The Fine Art of Propaganda: A Study of Father Coughlin's Speeches* (New York: Harcourt, Brace, 1939).

Lee, Douglas, "What's on the Horizon," available online at http://www. firstamendmentcenter.org/press/information/horizon.aspx?topic=FOI_horizon [Accessed May 1, 2008].

Lee, Matthew, "Government Changing Terrorism Language," available online at http://www.time.com/time/nation/article/0,8599,1734909,00.html?xid=feed-cnn-topics.

Levy, Leonard W., The Emergence of a Free Press (New York: Oxford University Press,1985).

Li, Xigen and Ralph Izard, "9/11 Attack Coverage Reveals Similarities, Differences," Newspaper Research Journal, 24, 1 (Winter 2003): 204–219.

Lichter, Robert, Stanley Rothman, and Linda Lichter, The Media Elite (Bethesda, Mid.: Adler & Adler, 1986).

Liebes, Tamar and Anat First, "Framing the Palestinian-Israeli Conflict," in Pippa Norris, Montague Kern, and Marion Just, eds. Framing Terrorism: The News Media, the Government, and the Public (New York: Routledge, 2003), 59–74.

Lippmann, Walter, Public Opinion (New York: Macmillan, 1922).

Livingston, Steven, "Clarifying the CNN Effect: An Examination of Media Effects According to Type of Military Intervention." John F. Kennedy School of Government's Joan Shorenstein Center on the Press, Politics and Public Policy at Harvard University, June 1997.

Luther, Catherine A. and Xiang Zhou, "Within the Boundaries of Politics: News Framing of SARS in China and the United States," Journalism & Mass Communication Quarterly, 82, 4 (2005): 857–872.

MacArthur, John, "Everybody Wants to Be at Versailles," in Kristina Borjesson, ed. Feet to the Fire: The Media after 9/11 (Amherst, N.Y.: Prometheus Books, 2005), 92–122.

MacIntyre, Alasdair, "Is Patriotism a Virtue?" in R. Beiner, ed. Theorizing Citizenship, (New York: State University of New York Press, 1995), 209–228.

Maguire, Sean, "When Does Reuters Use the Word Terrorist or Terrorism?" Available online at http://blogs.reuters.com/blog/2007/06/13/when-does-reuters-use-the-word-terrorist-or-terrorism/ [Accessed March 14, 2008].

"Maher Tapes Final Episode of 'Politically Incorrect,'" USA Today online, June 29, 2002.

Maher, Bill, "When Can We Finally Be Funny Again?" Los Angeles Times, September 10, 2006. http://www.latimes.com/news/printedition/opinion/la-op-maher10sep10,1,3356446.story.

Marighella, Carlos, Minimanual of the Urban Guerilla, http://www.marxists.org/archive/marighella-carlos/1969/06/minimanual-urban-guerrilla/index.htm[Accessed February 6, 2008].

McBride, Kelly, "Did Powerful Image Present an Unbalanced View?" Posted April 10, 2003, http://www.poynter.org/column.asp?id=53&aid=29510.

McCabe, David, "Patriotic Gore, Again," in Igor Primoratz, *Patriotism* (Amherst: N.Y.: Humanity Books, 2002), 121–141.

McClung Lee, Alfred and Elizabeth Briant Lee, eds., *The Fine Art of Propaganda: A Study of Father Coughlin's Speeches* (New York: Harcourt, Brace, 1939).

McCombs, Maxwell E. and Donald L. Shaw, "The Agenda-Setting Function of Mass Media," *Public Opinion Quarterly*, 36 (1972): 176-187.

McCombs, Maxwell E. and Amy Reynolds, "News Influence on Our Pictures of the World," in Jennings Bryant and Dolf Zillmann, eds. Media Effects (2nd ed.) (Mahwah, N.J.: Lawrence Erlbaum Associates, 2002), 1–18.

McLuhan, Marshall, *Understanding Media: The Extensions of Man* (New York: McGraw Hill, 1964).

McLuhan, Marshall and Bruce R. Powers, *The Global Village: Transformations in World Life and Media in the 21st Century* (Oxford: Oxford University Press, 1989).

McQuail, Denis, *Mass Communication Theory: An Introduction* (London: Sage, 1994).

———, *Sociology of Mass Communications* (Middlesex: Penguin Books, 1972).

Meiklejohn, Alexander, *Free Speech and Its Relation to Self-Government* (New York: Kennikat Press, 1948).

Merari, Ariel, "Terrorism as a Strategy of Insurgency," in Gérard Chaliand and Arnaud Blin, eds. *The History of Terrorism: From Antiquity to Al Qaeda* (Berkeley: University of California Press, 2007), 12-51.

Mermin, Jonathan, *Debating War and Peace: Media Coverage of U.S. Intervention in the Post–Vietnam Era* (Princeton: Princeton University Press, 1999).

Messaris, Paul, *Visual Persuasion: The Role of Images in Advertising* (Berkeley: Sage, 1997).

Messaris, Paul and Linus Abraham, "The Role of Images in Framing News Stories," in Stephen D. Reese, Oscar H. Gandy, Jr., and August E. Grant, eds. *Framing Public Life: Perspectives on Media and Our Understanding of the Social World* (Mahwah, N.J.: Lawrence Erlbaum Associates, 2001), 215-226.

Migaux, Philippe, "The Roots of Islamic Radicalism," in Gérard Chaliand and Arnaud Blin, eds. *The History of Terrorism: From Antiquity to Al Qaeda* (Berkeley: University of California Press, 2007), 255–313.

Miller, Abraham, *Terrorism, the Media and the Law* (New York: Transnational, 1982).
Miracle, Tammy, "The Army and Embedded Media," *Military Review* (September–October 2003), available online at http://findarticles.com/p/articles/mi_m0PBZ/is_5_83/ai_111573648 [Accessed May 1, 2008].

Moeller, Susan D., *Shooting War: Photography and the American Experience of Combat* (New York: Basic Books, 1989).

Morgan, David, "The Assassins: A Terror Cult," in the *International Encyclopedia of Terrorism* (Chicago: Fitzroy Dearborn, 1997), 40–41.

———, "Mongol Terror," in the *International Encyclopedia of Terrorism* (Chicago: Fitzroy Dearborn, 1997), 42–43.

Morley, Henry, ed., *Famous Pamphlets: Milton's Areopagitica, Killing No Murder, De Foe's Shortest Way with Dissenters, Steele's Crisis, Whatley's Historic Doubts Concerning Napoleon Bonaparte, Copleston's Advice to a Young Reviewer and Morley's Universal Library* 43 (New York: George Routledge and Sons, 1886).

Mueller, John E., *War, Presidents and Public Opinion* (New York: John Wiley and Sons, 1973)

Mundorf, Norbert, Dan Drew, Dolf Zillmann, James Weaver, "Effects of Disturbing News on Recall of Subsequently Presented News," *Communication Research*, 17 (1990): 601.

Murray, David, Joel Schwartz, S. Robert Lichter, *It Ain't Necessarily So: How Media Make and Unmake the Scientific Picture of Reality* (Lanham, Mid.: Rowman & Littlefield, April 25, 2001).

Nacos, Brigitte L., *Mass-Mediated Terrorism: The Central Role of the Media in Terrorism and Counterterrorism* (Lanham, Mid.: Rowman & Littlefield, 2007).

———, *Terrorism & the Media: From the Iranian Hostage Crisis to the Oklahoma City Bombing* (New York: Columbia University Press, 1996).

———, "Terrorism as Breaking News: Attack on America," *Political Science Quarterly* (Spring 2003): 23–52.

Nacos, Brigitte L. and Oscar Torres-Reyna, *Fueling Our Fears: Stereotyping, Media Coverage, and Public Opinion of Muslim Americans*, (Lanham, Mid.: Rowman & Littlefield, 2006).

Nacos, Brigitte L, Yaeli Bloch-Elkon, and Robert Y. Shapiro,"Post-9/11 Terrorism Threats, News Coverage, and Public Perceptions in the United States," *International Journal of Conflict and Violence*, 1 (2007): 105.

Nagar, Na'ama, "Frames that Don't Spill: The News Media and the War on Terrorism" (Paper presented at the International Studies Association [ISA] Annual Convention, Chicago, Illinois, February 24, 2007).

National Commission on Terrorist Attacks, *The 9/11 Commission Report: Final Report of the National Commission on Terrorist Attacks upon the United States* (New York: W.W. Norton, 2004), 20.

National Public Radio, "Roundtable: Egypt Terrorism Attacks, AFL-CIO Troubles," aired July 25, 2005. http://www.npr.org/templates/story/story.php?storyId=4769688.

Netanyahu, Benjamin, *International Terrorism: Challenge and Response* (Edison, N.J.: Transaction, 1982).

Neuman, W. Russell, Marion R. Just, and Ann N. Crigler, *Common Knowledge: News and the Construction of Political Meaning* (Chicago: University of Chicago Press, 1992).

Newhagen, John E., "TV News Images That Induce Anger, Fear, and Disgust: Effects on Approach-Avoidance and Memory," *Journal of Broadcasting & Electronic Media*, 42 (1998): 265–277.

Newhagen, John E. and Byron Reeves, "The Evening's Bad News: Effects of Compelling Negative Television News Images on Memory," *Journal of Communication*, 42 (1992): 25–41.

Nichols, John and Robert W. McChesney, *Tragedy and Farce: How the American Media Sell Wars, Spin Elections, and Destroy Democracy* (New York: New Press, 2005).

Norris, Pippa, Montague Kern, and Marion Just, eds., *Framing Terrorism: The News Media, the Government, and the Public* (New York: Routledge, 2003).

Nossek, Hillel and Daniel Berkowitz, "Telling 'Our' Story through News of Terrorism: Mythical Newswork as Journalistic Practice in Crisis," *Journalism Studies*, 7, 5 (2006): 691.

Nussbaum, Martha, "Patriotism and Cosmopolitanism," *Boston Review*, 19, 5 (Fall 1994). Retrieved online at http://www.soci.niu.edu/~phildept/Kapitan/nussbaum1.html.

Oates, Sarah, "Comparative Aspects of Terrorism Coverage: Television and Voters in the 2004 U.S. and 2005 British Elections" (Paper presented at the Political Communication Section pre-APSA Conference, the Annenberg School for Communication, University of Pennsylvania, August 2006).

Oglesby, Christy, "Rape Victims' Names Withheld by Choice, Not Law: Statutes on Confidentiality Don't Trump Media's Constitutional Rights," CNN Web site October 16, 2003, http://www.cnn.com/2003/LAW/10/16/rape.confidential/.

O'Heffernan, Patrick, "A Mutual Exploitation Model of Media Influence in US Foreign Policy," in W Lance Bennett, ed. *Taken by Storm: The Media, Public Opinion and US Foreign Policy in the Gulf War* (Chicago: University of Chicago Press, 1994).

Östgaard, Einar, "Factors Influencing the Flow of News," *Journal of Peace Research*, 2,1 (1965): 39–63.

Otte, Thomas G, "Russian Anarchist Terror," in the *International Encyclopedia of Terrorism* (Chicago: Fitzroy Dearborn, 1997), 56–57.

Overholser, Geneva, "On Behalf of Journalism: A Manifesto for Change," 2006. Retrieved online at http://www.annenbergpublicpolicycenter.org/Overholser/20061011_JournStudy.pdf.

Paivio, Allan, *Mental Representation: A Dual Coding Approach* (Oxford: Oxford University Press, 1986).
Palast, Greg, "Palast Charged with Journalism in the First Degree," available online at http://www.gregpalast.com/palast-charged-with-journalism-in-the-first-degree/ [Accessed May 1, 2008].

Paletz, David L., John Z. Ayanian, and Peter A. Fozzard, "Terrorism on TV News: The IRA, the FALN and the Red Brigades," in William C. Adams, ed. *Television Coverage of International Affairs* (Norwood, N.J.: Ablex, 1982), 143–166.

Paletz, David L. and Alex P. Schmid, eds., *Terrorism and the Media* (Newbury Park, Calif.: Sage, 1992).

Palmer, Nancy, ed., *Terrorism, War and the Press* (Cambridge, Mass.: Joan Shorenstein Center on the Press, Politics and Public Policy, 2003).

Pearce, Fred, "Forum: As Seen on Television/Where Are His Experts Are Coming From," *New Scientist*, June, 16 1990.

Pew Research Center for the People and the Press, "Internet News Audience Highly Critical of News Organizations: Views of Press Values and Performance: 1985–2007," Released August 9, 2007.

———, "News Audiences Increasingly Politicized, "Online News Audience Larger, More Diverse," Released June 8, 2004.

———, "Public More Critical of Press, But Goodwill Persists: Online Newspaper Readership Countering Print Losses," Released June 26, 2005.

———, "Republicans Uncertain on Rove Resignation: Plurality Favors Centrist Court Nominee," July 19, 2005.

———, "Who Flies the Flag? Not Always Who You Might Think: A Closer Look at Patriotism," June 27, 2007.

Picard, Robert G., *Media Portrayals of Terrorism: Functions and Meaning of News Coverage* (Iowa: Iowa State University Press, 1993).

Picard, Robert G. and Paul D. Adams, "Characterization of Acts and Perpetrators of Political Violence in Three Elite U.S. Daily Newspapers," *Political Communication and Persuasion*, 4 (1987): 1–9.

Pincus, Walter, "Guerilla at the Washington Post," in Kristina Borjesson, ed. *Feet to the Fire: The Media after 9/11* (Amherst, N.Y.: Prometheus Books, 2005), 218–248, 221.

Pintak, Lawrence, "America's Media Bubble: A Willful Blindness," *International Herald Tribune*, November 19, 2006. http://www.iht.com/articles/2006/11/19/opinion/edpintak.php.

Post, Jerrold, "The Psychological Dynamics of Terrorism," in Louise Richardson, ed. *The Roots of Terrorism* (New York: Routledge, 2006): 17–28.

Potter, Stewart, "Or of the Press," *Hastings Law Review*, 26 (1976): 631–636.

Project for Excellence in Journalism, "Before and After: How the War on Terrorism Has Changed the News Agenda," November 19, 2001 http://www.journalism.org/node/289.

————, "How 9-11 Changed the Evening News," Released September 11, 2006. http://www.journalism.org/node/1839.

————, "2007 State of the News Media Annual Report on American Journalism," http://www.stateofthenewsmedia.org/2007/printable_localtv_publicattitudes.asp

Protected Critical Infrastructure Information (PCII) Program, available online at http://www.dhs.gov/xinfoshare/programs/editorial_0404.shtm [Accessed May 1, 2008].

Red Lion Broadcasting co. v. FCC, 395 U.S. 367 (1969).

Reese, Stephen D., "Understanding the Global Journalist: A Hierarchy-of-Influences Approach," Journalism Studies, 2 (2001): 173–187.

Reese, Stephen D., Oscar H. Gandy, Jr., and August E. Grant, eds., Framing Public Life: Perspectives on Media and Our Understanding of the Social World (Mahwah, N.J.: Lawrence Erlbaum and Associates, 2001).

Reich, Walter, ed., Origins of Terrorism: Psychologies, Ideologies, Theologies, States of Mind (Washington, D.C.: Woodrow Wilson Center Press, 1990).

Reynolds, Amy, "How 'Live' Television Coverage Affects Content: A Proposed Model of Influence and Effects" (Paper presented at the International Communication Association Conference Mass Communication Division, Montreal, Quebec, May 23, 1997).

Reynolds, Amy and Brooke Barnett, "America under Attack" CNN's Visual and Verbal Framing of September 11," in Steven Chermak, Frank Bailey, and Michelle Brown, eds. Media Representations of September 11th (New York: Praeger, 2003), 85–101.

————, "This Just in … How National TV News Handled the Breaking Live Coverage of September 11th," Journalism & Mass Communication Quarterly, 80 (2003): 689–703.

Richardson, Louise, ed., The Roots of Terrorism (New York: Routledge, 2006).

Richmond Newspapers v. Virginia, 448 U.S. 555 (1980).

Rivers, William, Wilbur Schramm and Clifford Christians, Responsibility in Mass Communication (New York: Harper and Row, 1980).

Rutenberg, Jim and Bill Carter, "Draping Newscasts with the Flag," New York Times, September 20, 2001.

Ryan, Michael, "Framing the War against Terrorism," International Communication Gazette, 66, 5 (2004): 363–382.

Saloman, Gavriel, Interaction of Media, Cognition, and Learning (San Francisco: Josey-Bass, 1979).

Samuelson, Robert, "Unwitting Accomplices," Washington Post, November 7, 2001, A29.

Sanchez-Cuenca, Ignacio, "The Causes of Revolutionary Terrorism," in Louise Richardson, ed. *The Roots of Terrorism* (New York: Routledge, 2006), 71–82.

Schaefer, Todd M., "Framing the US Embassy Bombings and September 11 Attacks in African and U.S. Newspapers," in Pippa Norris, Montague Kern, and Marion Just, eds. *Framing Terrorism: The News Media, the Government, and the Public* (New York: Routledge, 2003), 93–112.

Schaffert, Richard, *Media Coverage and Political Terrorists: A Quantitative Analysis* (New York: Praeger, 1992).

Schechter, Danny, *Media Wars: News at a Time of Terror* (Lanham, Mid.: Rowman & Littlefield, 2003).

Schlesinger, Philip, Graham Murdock, and Philip Elliott, *Televising Terrorism: Political Violence in Popular Culture* (London: Comedia/Marion Boyars. 1983).

Schmid, Alex P., "The Problems of Defining Terrorism," in the *International Encyclopedia of Terrorism* (Chicago: Fitzroy Dearborn, 1997), 11–21.

Schmid, Alex P. and Janny de Graaf, *Violence as Communication: Insurgent Terrorism and the Western News Media* (London: Sage, 1982).

Schmid, Alex P. and Albert Jongman, *Political Terrorism: A New Guide to Actors, Authors, Concepts, Data Bases, Theories and Literature* (New Brunswick, N.J.: Transaction Books, 1988).

Schmidt, Steve, "Rumsfeld Response Due on Embedded Reporters; CNN Sniper Video Criticized, Praised," *San Diego Union-Tribune*, October 28, 2006.

Sebti, Bassam, "Heading into Danger," available online at http://cpj.org/Briefings/2006/DA_spring_06/bassam/bassam_DA.html [Accessed April 17, 2008].

Seib, Philip, *Beyond the Front Lines: How the News Media Cover a World Shaped by War* (New York: Palgrave Macmillan, 2004), 79.

———, "The News Media and the "Clash of Civilizations," *Parameters*, 34 (2004): 71–85.

Semetko, Holli A. and Patti M. Valkenburg, "Framing European Politics: A Content Analysis of Press and Television News," *Journal of Communication*, 50 (2000): 93–109.

Severin, Werner J. and James W. Tankard, Jr., *Communication Theories: Origins, Methods and Uses in the Mass Media* (3rd. ed.) (New York: Longman, 1992).

Shales, Tom, "A Media Role in Selling the War? No Question," *Washington Post*, April 25, 2007.

Shambaugh, George and William Josiger, "Fear Factor: The Impact of Terrorism on Public Opinion" (Paper presented at the ISA Annual Convention, Chicago, Illinois, February, 2007).

Shaw, David, "A Skeptical Journalist Isn't an Unpatriotic One," *Los Angeles Times* (April 20, 2003): E16.

Shoemaker, Pam and Stephen D. Reese, *Mediating the Message: Theories of Influence on Mass Media Content* (New York: Longman, 1996).

Sigal, Leon V., "Sources Make the News," in Robert Manoff and Michael Schudson, eds. *Reading the News* (New York: Pantheon, 1986), 9–37.

Silvio, Waisbord, "Journalism Risk and Patriotism," in Barbie Zelizer and Stuart Allan, eds. *Journalism after September 11* (London: Routledge, 2002), 201–216.

Simon, Jeffrey D., *The Terrorist Trap: America's Experience with Terrorism.* (Bloomington, Ind.: Indiana University Press, 2001).

Singh, Ramindar, "Covering September 11 and Its Consequences: A Comparative Study of the Press in America, India and Pakistan," in Nancy Palmer, ed. *Terrorism, War and the Press* (Cambridge, Mass.: Joan Shorenstein Center on the Press, Politics and Public Policy, 2003), 43.

Slone, Michelle, "Responses to Media Coverage of Terrorism," *Journal of Conflict Resolution,* 44 (2000): 522.

Smith, Jeffery A., *Printers and Press Freedom: The Ideology of Early American Journalism* (New York: Oxford University Press, 1988).

Smith, Tom W. and Lars Jarkko, *National Pride in Cross-national Perspective* (Chicago, Ill.: University of Chicago National Opinion Research Center, April 2001).

Smolkin, Rachel, "Too Transparent?" *American Journalism Review* (April/May 2006).

Soley, Lawrence D., "Pundits in Print: "Experts" and Their Use in Newspaper Stories," *Newspaper Research Journal,* 15 (spring 1994): 65–75.

Stempel III, Guido H. and Thomas Hargrove, "From an Academic: Newspapers Played Major Role in Terrorism Coverage," *Newspaper Research Journal,* 24, 1 (Winter 2003): 55–57.

Stone, David and David Hartley, *Media Ethics* (Princeton, N.J.: Films for the Humanities & Sciences, 1998).

Strossen, Nadine, *Forum on National Security and the Constitution,* available online at www.aclu.org/congress/1012402a.html [Accessed May 1, 2008].

Strugatch, Warren, "When Patriotism and Journalism Clash," *New York Times,* October 7, 2001.

Surette, Ray, *Media, Crime and Criminal Justice: Images and Reality* (Belmont, Calif.: Wadsworth, 1998).

Swain, Kristen Alley, "Outrage Factors and Explanations in News Coverage of the Anthrax Attacks," *Journalism and Mass Communication Quarterly,* 84, 2 (2007): 347.

Tankard, Jr. James W., "The Empirical Approach to the Study of Media Framing," in Stephen D. Reese, Oscar H. Gandy, Jr., and August E. Grant, eds. *Framing Public Life: Perspectives on Media and Our Understanding of the Social World* (Mahwah, N.J.: Lawrence Erlbaum Associates, 2001), 95–106.

Tankard, Jr. James W., Laura Hendrickson, J. Silberman, K. Bliss, and Salma Ghanem, "Media Frames: Approaches to Conceptualization and Measurement" (Paper presented at the annual meeting of the Association for Education in Journalism and Mass Communication, Boston, Mass., 1991).

Thomas, Helen, "Grande Dame: Persona Non Grata," in Kristina Borjesson, ed. *Feet to the Fire: The Media after 9/11* (Amherst, N.Y.: Prometheus Books, 2005), 323.

———, *Watchdogs of Democracy?: The Waning Washington Press Corps and How It Has Failed the Public* (New York: Charles Scribner's Sons, 2006).

Tierney, Kathleen, Christine Bevc, and Erica Kuligowski, "Metaphors Matter: Disaster Myths, Media Frames, and Their Consequences in Hurricane Katrina," *Annals of the American Academy of Political and Social Science*, 604, 1 (2006): 57–81.

Tolstoy, Leo, "Patriotism and Government," retrieved online at http://dwardmac.pitzer.edu/ Anarchist_Archives/bright/tolstoy/patriotismandgovt.html.

Tremlett, J. Edward, "'Politically Incorrect' Cancelled—Was It ABC or APC?" *The American Partisan*, May 15, 2002. http://www.american-partisan.com/cols/2002/tremlett/qtr2/0515.htm.

Trenchard, John and Thomas Gordon, eds., *Essays on Liberty, Civil and Religious, and Other Important Subjects* (Indianapolis: Liberty Fund, 1995).

Tuchman, Gaye, "Objectivity as Strategic Ritual: An Examination of Newsmen's Notions of Objectivity," *American Journal of Sociology*, 77 (1977): 660–679.

Turk, Judy VanSlyke, "Between President and Press: White House Public Information and Its Influence on the News Media" (Paper presented at the Association for Education in Journalism and Mass Communication, August, 1987, San Antonio, Texas).

The U.S. Army & Marine Corps Counterinsurgency Field Manual (U.S. Army Field Manual No. 3-24, Marine Corps Warfighting Publication No. 3-33.5) (Chicago: University of Chicago Press, 2007).

U.S. Department of State, *Patterns of Global Terrorism*, v., cited in Ariel Merari, "Terrorism as a Strategy of Insurgency," in Gérard Chaliand and Arnaud Blin, eds., *The History of Terrorism: From Antiquity to Al Qaeda* (Berkeley: University of California Press, 2007), 12–51.

5 U.S.C. 552 (b)(6-7) (C) (1988).

Walmsley, Jane, *Brit-Think Ameri-Think* (New York: Penguin, 1986).

Walzer, Michael, *Just and Unjust Wars: A Moral Argument with Historical Illustrations* (New York: Basic Books, 1977).

Wanta, Wayne, Mary Ann Stephenson, Judy VanSlyke Turk, and Maxwell E. McCombs, "How President's State of Union Talk Influenced News Media Agendas," *Journalism Quarterly* 66,3 (1989): 537–541.

Weaver, David H. "Political Issues and Voter Need for Orientation," in Donald Shaw and Maxwell McCombs, eds. The Emergence of American Political Issues (St. Paul, Minn.: West, 1977), 107–119.

Weaver, David H., Randal A. Beam, Bonnie J. Brownlee, Paul S. Voakes, and G. Cleveland Wilhoit, The American Journalist in the 21st Century: U.S. News People at the Dawn of a New Millennium (Mahwah, N.J.: Lawrence Erlbaum Associates, 2007).

Weaver, David H., Doris A. Graber, Maxwell E. McCombs, and Chaim H. Eyal, Media Agenda-Setting in a Presidential Election: Issues, Images, and Interest (New York: Praeger, 1981).

Weaver, David H. and G. Cleveland Wilhoit, The American Journalist: A Portrait of U.S. News People and Their Work (2nd. ed.) (Bloomington, Ind.: Indiana University Press, 1991).

———, The American Journalist in the 1990s: U.S. News People at the End of an Era (Mahwah, N.J.: Lawrence Erlbaum Associates, 1996).

Weimann, Gabrielle and Conrad Winn, The Theater of Terror: The Mass Media and International Terrorism (New York: Longman/Addison-Wesley, 1994).

Wells, Matt, "World Service Will Not Call U.S. Attacks Terrorism," available online at http://www.guardian.co.uk/media/2001/nov/15/warinafghanistan2001.afghanistan [Accessed March 15, 2008]. .

White, Jonathan R., Terrorism: An Introduction (3rd ed.) (Belmont, Calif.: Wadsworth, 2002).

Williams, Raymond, Marxism and Literature (New York: Oxford University Press, 1977).

Worawongs, Worapron, Weirui Wang, and Ashley Sims, "U.S Media Coverage of Natural Disasters: A Framing Analysis of Hurricane Katrina and the Tsunami" (Paper presented at the annual meeting of the Association for Education in Journalism and Mass Communication, August 8, 2007).

Yellin, Tom, "Inside the Ratings Vise," in Kristina Borjesson, ed. Feet to the Fire: The Media after 9/11 (Amherst, N.Y.: Prometheus Books, 2005).

Zinn, Howard, Declarations of Independence: Cross-Examining American Ideology (New York: Harper Perennial, 1990).

Ziomek, Jon, Journalism, Transparency and the Public Trust: A Report of the Eighth Annual Aspen Institute Conference on Journalism and Society, 2005.

INDEX

ABOUT THE AUTHORS

Brooke Barnett is Associate Professor in the School of Communications at Elon University and Director of the Elon Program for Documentary Production. She has a Ph.D. in mass communication from Indiana University. She is co-editor of *Multidisciplinary Approaches to Communication Law Research* (2006) and author of *Media Coverage of Crisis: The War on Terror and the Wars in Iraq* (2005) as well as numerous journal articles and book chapters. Her professional background includes work as a news director, documentary producer, reporter, and producer in public television.

Amy Reynolds is Associate Professor and Associate Dean for Research and Graduate Studies in the School of Journalism at Indiana University. She has a Ph.D. in mass communication from the University of Texas at Austin. She is the author or editor of four books and numerous refereed journal articles and book chapters. Her professional background includes work as a reporter and editor at newspapers and as a reporter, producer, and news director at local television news stations.

Mediating
American
History

SERIES EDITOR: DAVID COPELAND

Realizing the important role that the media have played in American history, this series provides a venue for a diverse range of works that deal with the mass media and its relationship to society. This new series is aimed at both scholars and students. New book proposals are welcomed.

For additional information about this series or for the submission of manuscripts, please contact:

Mary Savigar, Acquisitions Editor
Peter Lang Publishing, Inc.
29 Broadway, 18th floor
New York, New York 10006
Mary.Savigar@plang.com

To order other books in this series, please contact our Customer Service Department:

(800) 770-LANG (within the U.S.)
(212) 647-7706 (outside the U.S.)
(212) 647-7707 FAX

Or browse by series:

WWW.PETERLANG.COM